Preface

This Study Guide has been designed to accompany the tenth edition of *Nutrition: Concepts and Controversies* and its chapter divisions parallel those in the text. The primary goal of the Study Guide is to facilitate your study of the material presented in the text and enhance your learning. Pages in the Study Guide have been perforated and three-hole punched to enable you to insert, delete, or rearrange pages to best suit your needs or to correspond with the sequence used by your instructor.

Each chapter consists of seven sections: chapter objectives, key concepts, a fill-in-the-blank chapter summary, chapter glossary of key terms, exercises which include chapter study questions, short answer questions, problem-solving/application questions and controversy questions, sample test items, and an answer key. In addition, some chapters have study aids that complement material presented in the text.

To obtain the maximum benefit from this Study Guide you should: (1) read the chapter in the text; (2) study your class notes, and (3) complete the corresponding chapter in the Study Guide, including the following sections:

- **Chapter Objectives and Key Concepts.** Read the objectives and key concepts at the beginning of each chapter to help you focus your attention on the most essential material and concepts presented in the chapter.

- **Summing Up.** This section provides a brief summary of the chapter using a fill-in-the-blank format. To help you summarize and review the main points in the chapter, read the summary and fill in the blanks.

- **Chapter Glossary.** This section includes key terms and their definitions presented in the margins of the text. Match the definitions and terms or complete the crossword puzzles to enhance your understanding of the material and prepare for many of the test items.

- **Exercises.** Chapter study questions, short answer questions, problem-solving/application questions and controversy questions are included in this section. Chapter study questions and controversy questions allow you to review key concepts and help you prepare for essay questions on examinations, while short answer and problem-solving/application questions help you to review and apply key concepts and principles presented.

- **Study Aids.** Many chapters include study aids to help you remember, understand and practice key information presented in the text. Some study aids refer to tables and figures in your textbook.

- **Sample Test Items.** Comprehension level and application level multiple choice, matching, and true/false items are included to help you test your recall and understanding of the basic content presented in the text. Many of the questions are likely to be similar to those included on examinations. However, rather than simply learning the answers to the sample test questions, you should use the questions in this section as the basis for learning and studying important concepts and principles presented in the chapter.

- **Answers.** Answers are provided for all sections of the Study Guide. When checking fill-in-the-blank answers, be aware that more than one word may satisfactorily complete a sentence. If your answer differs from the one provided, determine if you have used an acceptable synonym or whether you missed the basic concept. When checking chapter study questions and controversy questions, keep in mind that only key points have been included in the answer. Good answers to discussion questions require not only a recall of the key points but also descriptions of interrelationships and explanations of why and how. A well-written answer to discussion questions requires synthesis of knowledge gained from previous lessons, related courses, and personal experience. After completing the sections of the chapter, check your answers and use your results to reinforce and strengthen the learning process by going back and restudying areas that you were unsure of or missed.

We hope you find this Study Guide to be useful. Please let us hear from you if you have suggestions for ways to improve the Study Guide.

Jana Kicklighter, Ph.D., R.D.
Department of Nutrition
Georgia State University
Atlanta, Georgia 30303

Study Guide

for

Sizer and Whitney's

Nutrition
Concepts and Controversies

Tenth Edition

Jana R. Kicklighter
Georgia State University

THOMSON

WADSWORTH

Australia • Canada • Mexico • Singapore • Spain • United Kingdom • United States

Printer: Thomson/West

0-534-40674-2

Thomson Higher Education
10 Davis Drive
Belmont, CA 94002-3098
USA

Table of Contents

Chapter 1 - Food Choices and Human Health

Chapter Objectives

After completing this chapter, you should be able to:

1. Differentiate between those diseases that are strongly influenced by diet and those that are not responsive to nutrition.

2. Identify the benefits of physical activity.

3. List the six classes of nutrients including which are organic and which are energy-yielding.

4. Describe the *Dietary Guidelines for Americans* and *Healthy People 2010.*

5. Describe the process by which scientists uncover nutrition facts.

6. Describe the cultural and social meanings attached to food.

7. Identify and explain the five characteristics of a nutritious diet.

8. Begin planning a healthy diet by using the concept of nutrient density.

9. Distinguish valid nutrition information from misinformation (Controversy 1).

Key Concepts

✓ The nutrients in food support growth, maintenance, and repair of the body. Deficiencies, excesses, and imbalances of nutrients bring on the diseases of malnutrition.

✓ Nutrition profoundly affects health.

✓ Choice of diet influences long-term health within the range set by genetic inheritance. Nutrition has little influence on some diseases but strongly affects others.

✓ Nutritional genomics holds great promise for advances in nutrition science.

✓ Personal life choices, such as staying physically active or using tobacco or alcohol, also affect health for the better or worse.

✓ Food supplies energy and nutrients. Foremost among the nutrients is water. The energy-yielding nutrients are carbohydrates, fats (lipids), and protein. The regulator nutrients are vitamins and minerals. Food energy is measured in calories; food and nutrient quantities are often measured in grams.

✓ In addition to nutrients, food conveys emotional satisfaction and hormonal stimuli that contribute to health. Foods also contain phytochemicals that give them their tastes, aromas, colors, and other characteristics. Some phytochemicals may play roles in reducing disease risks.

✓ Foods come in a bewildering variety in the marketplace, but the foods that form the basis of a nutritious diet are basic foods, such as ordinary milk and milk products; meats, fish, and poultry; vegetables and dried peas and beans; fruits; and grains.

✓ A well-planned diet is adequate in nutrients, is balanced with regard to food types, offers food energy that matches energy expended in activity, is moderate in unwanted constituents, and offers a variety of nutritious foods.

✓ Cultural traditions and social values revolve around food. Some values are expressed through foodways. Many factors other than nutrition drive food choices.

✓ Scientists uncover nutrition facts by experimenting. Single studies must be replicated before their findings can be considered valid. New nutrition news is not always to be believed; established nutrition news has stood up to the test of time.

✓ The *Dietary Guidelines for Americans, Nutrition Recommendations for Canadians*, and other recommendations address the problems of overnutrition and undernutrition. To implement them requires exercising regularly; following the USDA Food Guide; seeking out whole grains, fruits, and vegetables, healthful fats; and limiting intakes of saturated fats, sugar, and salt, and moderating alcohol intake.

✓ The U.S. Department of Health and Human Services sets nutrition objectives for the nation each decade.

Summing Up

Nutrition is the study of how (1)_____ nourishes the body. The human body uses

(2)_____ to move and work and it comes indirectly from the sun by way of plants. The body

also requires (3)_____ kinds of nutrients. Four of these six are (4)_____, or

carbon containing. Foremost among the six classes of nutrients in foods is (5)_____. Three of the

nutrients are energy-yielding, including carbohydrates, (6)_____, and protein. The fifth and sixth

classes of nutrients are the vitamins and (7)_____, which act as regulators.

Some of the nutrients are (8)_____ nutrients, meaning that the body cannot make them for

itself. Dietary supplements are not appropriate substitutes for food because food offers more than just the six basic

(9)_____. For example, elemental diets can support life but not optimal

(10)_____ and health, and they can often lead to medical complications. In addition, food helps

maintain the intestine and conveys messages of comfort to the (11)_____. Many compounds

other than the six classes of nutrients are also present in foods and are referred to as (12)_____.

Scientists discover nutrition facts by conducting (13)_____. Single studies must be

(14)_____ before findings can be confirmed.

Among the many factors influencing food-related choices are (15)_____ traditions, which evolve

as people move about, learn about new foods, and teach each other. People from every country enjoy special

(16)_____ that represent their own heritage. Among factors people cite to explain daily food choices

are advertising, personal preference, habit, ethnic heritage, (17)_____ pressure, availability, convenience, economy, positive associations, emotional comfort, values or beliefs, region of the country, weight and (18)_____ value. A key to wise diet planning is to make sure that the foods you eat daily, your (19)_____ foods, are especially nutritious.

The U.S. Department of Health and Human Services sets ten-year health objectives for the nation in its document (20) _____ . Nutrition (21)_____ makes it possible for researchers to assess the nutrient status, health indicators, and dietary intakes of the U.S. population.

According to the *Dietary Guidelines for Americans*, a healthful diet is rich in a variety of fruits, vegetables, (22)_____ and nonfat or low-fat milk products. The extent to which foods support good health depends on the (23)_____, nutrients and nonnutrients they contain. A nutritious diet has five characteristics, including (24)_____, balance, calorie control, (25)_____, and variety.

Chapter Glossary

Matching Exercise:		
_____ 1. nutrition	a.	components of food that are indispensable to the body's functioning; they provide energy, serve as building material, help maintain or repair body parts, and support growth
_____ 2. essential nutrients		
_____ 3. food	b.	carbon containing
_____ 4. nonnutrients	c.	the study of the nutrients in foods and in the body
	d.	the foods (including beverages) a person eats and drinks
_____ 5. organic	e.	nonnutrient compounds in plant-derived foods having biological activity in the body
_____ 6. foodways	f.	the sum of a culture's habits, customs, beliefs, and preferences concerning food
_____ 7. energy-yielding nutrients		
_____ 8. nutrients	g.	foods associated with particular cultural subgroups within a population
_____ 9. energy	h.	the capacity to do work
_____ 10. legumes	i.	the nutrients the body cannot make for itself (or cannot make fast enough) from other raw materials
_____ 11. phytochemicals	j.	the nutrients the body can use for energy
_____ 12. dietary supplements	k.	pills, liquids, or powders that contain purified nutrients or other ingredients
_____ 13. ethnic foods	l.	medically, any substance that the body can take in and assimilate that will enable it to stay alive and to grow
_____ 14. diet	m.	beans, peas and lentils valued as inexpensive sources of protein, vitamins, minerals and fiber that contribute little or no fat to the diet
_____ 15. genome		
	n.	compounds other than the six nutrients that are present in foods and that have biological activity in the body
	o.	the full complement of genetic material in the chromosomes of a cell

Crossword Puzzle:

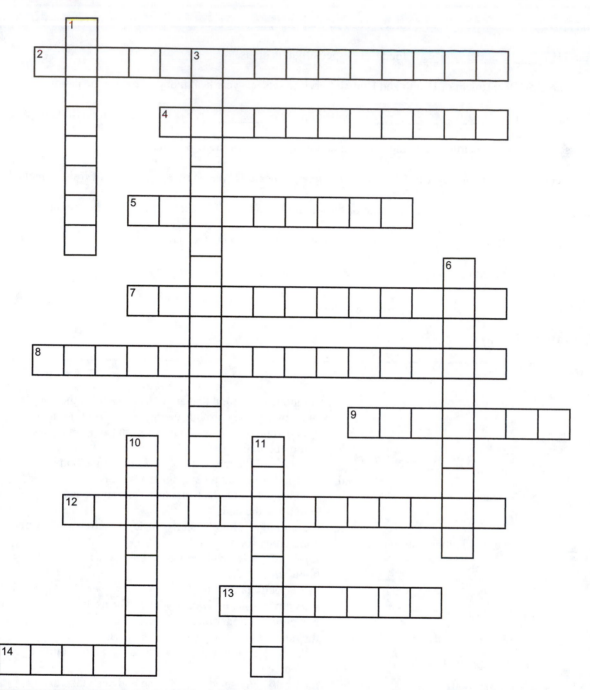

4

Across:	Down:
2. a measure of nutrients provided per calorie of food	1. styles of cooking
4. people who exclude from their diets animal flesh and possibly other animal products such as milk, cheese, and eggs	3. diets composed of purified ingredients of known chemical composition
5. people who eat foods of both plant and animal origin, including animal flesh	6. the dietary characteristic of providing constituents within set limits, not to excess
7. any condition caused by excess or deficient food energy or nutrient intake or by an imbalance of nutrients	10. units of energy
8. long-duration degenerative diseases characterized by deterioration of the body organs	11. the dietary characteristic of providing all of the essential nutrients, fiber, and energy in amounts sufficient to maintain health and body weight
9. the dietary characteristic of providing foods of a number of types in proportion to each other, such that foods rich in some nutrients do not crowd out of the diet foods that are rich in other nutrients	
12. control of energy intake, a feature of a sound diet plan	
13. the dietary characteristic of providing a wide selection of foods—the opposite of monotony	
14. units of weight	

Exercises

Answer these chapter study questions:

1. What is meant by the statement that protein does double duty?

2. What is meant by the term essential nutrients?

3. What does food provide in addition to nutrients?

4. Why do nutrition scientists agree that people should not eat the same foods day after day?

5. Why does the variety of foods available today make it more difficult, rather than easier, to plan nutritious diets?

Complete these short answer questions:

1. The four organic nutrients include:

   a.
   b.
   c.
   d.

2. The six kinds of nutrients include:

   a.
   b.
   c.
   d.
   e.
   f.

3. The energy-yielding nutrients are:

   a.

   b.

   c.

4. Factors people cite to explain food choices include:

   a.   h.
   b.   i.
   c.   j.
   d.   k.
   e.   l.
   f.   m.
   g.   n.

5. Five characteristics of a nutritious diet include:

   a.

   b.

   c.

   d.

   e.

Solve these problems:

There are many factors which people cite to explain their food choices. For questions 1-7, match the descriptive statements on the left with the reasons for the food choices, listed on the right. Each letter is used only once.

_____ 1.	I always drink coffee at morning break.	a. personal preference
_____ 2.	I usually microwave a frozen dinner after getting home from my night class.	b. habit
		c. social pressure
_____ 3.	All my friends are going out for pizza after the game.	d. availability
_____ 4.	I use dried beans as a substitute for meat because meat costs too much.	e. convenience
_____ 5.	I eat a lot of meat because I like the taste.	f. economy
_____ 6.	I usually choose chicken rather than beef as a meat because it is healthier.	g. nutritional value
_____ 7.	I have to eat high-fat foods because I eat in the employee cafeteria every day for lunch.	

Answer these controversy questions:

1. What type of screening is required before research findings are published in scientific journals?

2. What criterion would denote that a physician is well-qualified to speak on nutrition?

3. What is a diploma mill?

4. What are the qualifications of a dietetic technician?

5. Which type of dietetic subspecialties would you expect to see in a hospital setting?

Study Aid

Use Table 1-2 on page 5 in your textbook to familiarize yourself with the elements in the six classes of nutrients.

Sample Test Items

Comprehension Level Items:

1. A scientist's responsibilities include all of the following **except**:

    a. obtaining facts by systematically asking questions.
    b. conducting experiments to test for possible answers.
    c. submitting their findings to the news media.
    d. publishing the results of their work in scientific journals.

2. Which of the following nutrients is **not** organic?

    a. minerals
    b. vitamins
    c. protein
    d. carbohydrates
    e. fat

3. Foremost among the six classes of nutrients in foods is:

    a. fats.
    b. carbohydrates.
    c. vitamins.
    d. water.

4. Which of the following is **not** an energy-yielding nutrient?

    a. carbohydrates
    b. vitamins
    c. fats
    d. protein

5. All vitamins and minerals:

    a. provide energy to the body.
    b. serve as parts of body structures.
    c. serve as regulators.
    d. are organic.

6. Characteristics of essential nutrients include:

    a. the body can make them for itself.
    b. they are found in all six classes of nutrients.
    c. they must be received from foods.
    d. a and b
    e. b and c

7. Which of the following is characteristic of elemental diets?

    a. They are sufficient to enable people to thrive.
    b. They support optimal growth and health.
    c. They provide everything that foods provide.
    d. They are life-saving for people in the hospital who cannot eat ordinary food.

8. Within the range set by your genetic inheritance, the likelihood that you will develop chronic diseases is strongly influenced by your food choices.

   a. true
   b. false

9. Which of the following is the least nutrition-responsive disease?

   a. osteoporosis
   b. hypertension
   c. low birthweight
   d. sickle-cell anemia

10. Which of the following research designs is the most powerful in pinpointing the mechanisms by which nutrition acts because it shows the effects of treatments?

    a. case studies
    b. laboratory studies
    c. epidemiological studies
    d. correlation studies

11. Which of the following is **not** a government agency responsible for monitoring the nation's nutrition?

    a. U.S. Department of Health and Human Services
    b. U.S. Department of Agriculture
    c. Food and Drug Administration
    d. Centers for Disease Control and Prevention

12. The variety of foods available today may make it more difficult, rather than easier, to plan nutritious diets.

    a. true
    b. false

13. A key to wise diet planning is to make sure your _____ foods are especially nutritious.

    a. processed
    b. fast
    c. enriched
    d. staple

14. Foods composed of parts of whole foods, such as butter, sugar, or corn oil, are called _____ foods.

    a. enriched
    b. partitioned
    c. processed
    d. staple

15. The description of a diet that provides no unwanted constituents in excess is called:

    a. variety.
    b. calorie control.
    c. moderation.
    d. balance.

16. The key to evaluating an individual food is to:

    a.   judge it as either good or bad.
    b.   identify how it can reduce your chance of developing an illness.
    c.   judge its appropriate role in the context of all other food choices.
    d.   listen to what other people say about it.

17. Foods that are rich in nutrients relative to their energy contents are called foods with high _____.

    a.   nutrient density
    b.   calorie control
    c.   nutrient balance
    d.   dietary moderation

18. The foods that present the most nutrients per calorie are the:

    a.   meats.
    b.   vegetables.
    c.   fruits.
    d.   grains.

19. The most energy-rich of the nutrients is:

    a.   carbohydrate.
    b.   protein.
    c.   fat.
    d.   alcohol.

20. Of the leading causes of death, _____ are related to nutrition, and 1 to alcohol consumption.

    a.   2
    b.   3
    c.   4
    d.   5

21. Healthy people who eat a healthful diet do not need liquid formula diets and most need no dietary supplements.

    a.   true
    b.   false

22. When a person is fed through a vein:

    a.   the digestive organs grow larger.
    b.   the body's defenses against infections weaken.
    c.   the digestive organs weaken.
    d.   a and b
    e.   b and c

23. Which of the following is an ongoing national research project that gathers information from about 50,000 people using diet histories, physical examinations and measurements, and laboratory tests?

    a.   *Dietary Guidelines for Americans*
    b.   National Health and Nutrition Examination Survey
    c.   *Healthy People 2010*
    d.   Continuing Survey of Food Intakes by Individuals

24. Successes related to *Healthy People 2000* included all of the following **except**:

    a. reductions in the incidence of foodborne infections.
    b. reductions in the incidence of heart disease.
    c. reductions in cancer death rates.
    d. reductions in the number of overweight people.

Application Level Items:

25. You are explaining to a friend the characteristics of the most healthful diet based on the *Dietary Guidelines*. Which of the following would you include in your explanation?

    a. low in cholesterol
    b. low in *trans* fat
    c. moderate in saturated fat
    d. a and b
    e. a and c

26. One of your best friends makes the statement that eggs are a bad food. Which of the following would be the best response to this statement?

    a. Yes they are because they have a lot of cholesterol.
    b. No they are not because they provide high quality protein.
    c. Yes they are because they cause heart disease.
    d. They are neither good nor bad depending on how they are used in the total diet.

27. How many calories are contained in a food that has 20 grams of carbohydrate, 15 grams of protein, and 8 grams of fat?

    a. 145
    b. 195
    c. 212
    d. 245

28. A food contains 8 grams of fat and a total of 212 calories. What percentage of the calories in the food come from fat?

    a. 25%
    b. 34%
    c. 46%
    d. 52%

29. An individual consumes a total of 3000 calories and 100 grams of fat in a day. Does this person's diet meet the goal of 20-35% of calories from fat?

    a. yes
    b. no

30. Which of the following characteristics of a nutritious diet is illustrated by the statement that a certain amount of fiber in foods contributes to health of the digestive tract, but too much fiber leads to nutrient losses?

    a. moderation
    b. variety
    c. balance
    d. adequacy

Answers

Summing Up

(1) food; (2) energy; (3) six; (4) organic; (5) water; (6) fats; (7) minerals; (8) essential; (9) nutrients; (10) growth; (11) brain; (12) nonnutrients; (13) experiments; (14) duplicated; (15) cultural; (16) foodways; (17) social; (18) nutritional; (19) staple; (20) Healthy People; (21) monitoring; (22) whole grains; (23) calories; (24) adequacy; (25) moderation

Chapter Glossary

Matching Exercise: (1) c; (2) i; (3) l; (4) n; (5) b; (6) f; (7) j; (8) a; (9) h; (10) m; (11) e; (12) k; (13) g; (14) d; (15) o

Crossword Puzzle: (1) cuisines; (2) nutrient density; (3) elemental diets; (4) vegetarians; (5) omnivores; (6) moderation; (7) malnutrition; (8) chronic diseases; (9) balance; (10) calories; (11) adequacy; (12) calorie control; (13) variety; (14) grams

Exercises

Chapter Study Questions:

1. Protein can yield energy, but it also provides materials that form structures and working parts of body tissues.

2. Essential nutrients are those that the body cannot make for itself from other raw materials; they are the nutrients that must be obtained in food and are needed by the body; without them you will develop deficiencies.

3. Foods are thought to convey emotional satisfaction and hormonal stimuli that contribute to health; also they provide nonnutrients and other compounds that give them their tastes, aromas, colors, and other characteristics believed to affect health by reducing disease risks.

4. One reason is that some less well-known nutrients and some nonnutrients could be important to health and some foods may be better sources of these than others. Another reason is that a monotonous diet may deliver large amounts of toxins or contaminants. Variety also adds interest and can be a source of pleasure.

5. The food industry offers thousands of foods, many of which are processed mixtures of the basic ones, and some of which are constructed mostly from artificial ingredients.

Short Answer Questions:

1. (a) carbohydrate; (b) fat; (c) protein; (d) vitamins

2. (a) water; (b) carbohydrate; (c) fat; (d) protein; (e) vitamins; (f) minerals

3. (a) carbohydrates; (b) fats; (c) protein

4. (a) personal preference; (b) habit; (c) ethnic heritage; (d) social pressure; (e) availability; (f) convenience; (g) economy; (h) positive associations; (i) emotional comfort; (j) values or beliefs; (k) nutritional value; (l) region of the country; (m) weight; (n) advertising

5. (a) adequacy; (b) balance; (c) calorie control; (d) moderation; (e) variety

Problem-Solving:

(1) b; (2) e; (3) c; (4) f; (5) a; (6) g; (7) d

Controversy Questions:

1. The work must survive a screening review by peers of scientists.

2. Membership in the American Society for Clinical Nutrition.

3. A business that awards meaningless degrees to those who pay the fees, without requiring its students to meet educational standards.

4. Completion of a two-year academic degree from an accredited college or university and an approved dietetic technician program.

5. Administrative, clinical and nutrition support.

Sample Test Items

1. c (pp. 11-12)
2. a (p. 4-5)
3. d (p. 5)
4. b (p. 5)
5. c (p. 5)
6. e (p. 5)
7. d (p. 6)
8. a (p. 3-4)
9. d (p. 3)
10. b (p. 12)
11. c (p. 13)
12. a (p. 7)
13. d (p. 7)
14. b (p. 7)
15. c (p. 8)
16. c (p. 7)
17. a (p. 19)
18. b (p. 19)
19. c (p. 5)
20. c (p. 3)
21. a (p. 6)
22. e (p. 6)
23. b (p. 12)
24. d (p. 17)
25. d (p. 16)
26. d (p. 8-9)
27. c (p. 5)
28. b (p. 5)
29. a (p. 5)
30. a (p. 9)

Chapter 2 - Nutrition Tools
Standards and Guidelines

Chapter Objectives

After completing this chapter, you should be able to:

1. Describe the roles *Dietary Reference Intakes (DRI)* and *Daily Values* play in achieving adequate nutrient intakes.

2. Describe how *Dietary Reference Intake* values are established for nutrients.

3. Recognize that many nations and international groups have published sets of standards similar to the DRI.

4. List the American College of Sports Medicine's recommendations for physical activity.

5. Identify the components of the *Dietary Guidelines for Americans* and the USDA Food Guide and discuss how they are used to plan healthy diets.

6. Explain how exchange systems are useful to diet planners.

7. Discuss how consumers can use the ingredients list, nutrition facts panel, and health messages on food labels to make healthy food choices.

8. Describe possible actions of phytochemicals and functional foods and the evidence supporting their role in fighting certain diseases (Controversy 2).

Key Concepts

✓ The Dietary Reference Intakes (DRI) are nutrient intake standards set for people living in the United States and Canada. The Daily Values are U.S. standards used on food labels.

✓ The DRI provide nutrient intake goals for individuals, provide a set of standards for researchers and makers of public policy, establish tolerable upper limits for nutrients that can be toxic in excess, and take into account evidence from research on disease prevention. The DRI are composed of the RDA, AI, UL and EAR lists of values along with the AMDR ranges for energy-yielding nutrients.

✓ The DRI represent up-to-date, optimal, and safe nutrient intakes for healthy people in the United States and Canada.

✓ The DRI are based on scientific data and are designed to cover the needs of virtually all healthy people in the United States and Canada.

✓ Estimated Energy Requirements are energy intake recommendations predicted to maintain body weight and to discourage unhealthy weight gain.

✓ The Daily Values are standards used only on food labels to enable consumers to compare the nutrient values among foods.

✓ Many nations and groups issue recommendations for nutrient and energy intakes appropriate for specific groups of people.

- The USDA Food Guide specifies the amounts of foods from each group people need to consume to meet their nutrient requirements without exceeding their calorie allowances.

- The messages of the USDA Food Guide are conveyed to consumers to improve the health of the nation.

- The USDA Food Guide can be used with flexibility by people with different eating styles.

- People wishing to avoid overconsuming calories must pay attention to the sizes of their food servings.

- Exchange lists facilitate calorie control by providing an understanding of how much carbohydrate, fat, and protein are in each food group.

- The USDA Healthy Eating Index is a tool for assessing diet quality.

Summing Up

A committee of qualified nutrition experts from the United States and (1)_____ develops and publishes nutrient intake recommendations known as the (2)_____. Most people need to focus on only two kinds of DRI values: those that set nutrient intake goals for individuals, the Recommended Dietary Allowances and (3)_____; and those that define a safe upper limit of nutrient intake (UL). These recommendations are intended to be used to plan diets for (4)_____. These recommended intakes take into account (5)_____ as well as nutrient adequacy. Other values established by the DRI Committee include (6)_____, which are used by researchers and policy makers, and Tolerable Upper Intake Levels. The value for energy intake is set at the (7)_____ of the population's estimated energy requirements, and healthy ranges of intake for each energy nutrient, known as the (8)_____, have also been established. Many nations and international groups have published standards similar to the DRI but they (9)_____ from the DRI in some respects.

To help people plan menus, two kinds of planning tools are available including food group plans and the (10)_____ system. The USDA Food Guide is based on five major groups, some of which have (11)_____. The USDA Food Guide sorts foods in each group by (12)_____; (13)_____ are divided into 5 subgroups according to their nutrient contents. The concept of the (14)_____—the difference between the calories needed to maintain weight and those needed to supply nutrients—gives people the option to consume some less nutrient-dense foods while controlling (15)_____.

Exchange systems are useful for those wishing to control (16)_____, for those who must control carbohydrate intakes, and those who should control their intake of (17)_____ and saturated fat. The lists organize foods according to their calories, carbohydrate, fat, saturated fat and (18)_____ contents. The exchange system highlights a fact that the USDA Food Guide also points out, that most foods provide more than just one (19)_____ nutrient.

The Nutrition Education and Labeling Act of 1990 set the requirements for nutrition labeling information. According to law, every packaged food must state the (20)_____ name of the product, the name and address of the manufacturer, the (21)_____ contents and the ingredients, in (22)_____ order of predominance by weight. The (23)_____ informs consumers of the nutrient contents of the food. In addition to food energy, the label must include total fat (with breakdown of saturated fat and *trans* fat), cholesterol, sodium, total carbohydrate (including starch, fiber and sugars), (24)_____, vitamins A and C, and the minerals (25)_____ and iron. Some claims about health are allowed on food labels and these are regulated by the (26)_____.

Chapter Glossary

Matching Exercise:

_____ 1. requirement

_____ 2. discretionary calorie allowance

_____ 3. food group plans

_____ 4. Daily Values

_____ 5. Healthy Eating Index

_____ 6. balance study

_____ 7. exchange system

_____ 8. Estimated Energy Requirement

_____ 9. Dietary Reference Intakes

a. nutrient standards that are printed on food labels
b. the amount of a nutrient that will just prevent the development of specific deficiency signs
c. diet planning tools that sort foods into groups based on nutrient content and then specify that people should eat certain minimum numbers of servings of foods from each group
d. a laboratory study in which a person is fed a controlled diet and the intake and excretion of a nutrient are measured
e. a diet planning tool that organizes foods with respect to their nutrient contents and calorie amounts
f. a set of four lists of values for measuring the nutrient intakes of healthy people in the United States and Canada
g. the average dietary energy intake predicted to maintain energy balance in a healthy adult of a certain age, gender, weight, height, and level of physical activity consistent with good health
h. the balance of calories remaining in a person's energy allowance after accounting for the number of calories needed to meet recommended nutrient intakes through consumption of nutrient-dense foods
i. dietary assessment tool that evaluates a diet's adherence to the principles of the USDA Food Guide and the *Dietary Guidelines for Americans*, as well as the variety of foods the diet contains

Exercises

1. Why is the value set for energy intake not generous?

2. Describe the goals the DRI committee had in mind when setting the DRI values.

3. List the foods most Americans need to choose more often and less often in order to meet the ideals of the 2005 *Dietary Guidelines*.

4. Explain how the concept of the discretionary calorie allowance may be used to control intake of calories while meeting nutrient needs.

5. Explain why the DRI values are not used on food labels.

6. How do USDA Food Guide serving equivalents compare to the portion sizes served in restaurants?

Complete these short answer questions:

1. The exchange system lists estimates of these nutrients in foods:

   a.
   b.
   c.
   d.

2. The UL are indispensable to consumers who:

   a.
   b.

3. Dietary Reference Intakes include these five lists of values:

   a.
   b.
   c.
   d.
   e.

4. The five groups in the USDA Food Guide include:

   a.   d.
   b.   e.
   c.

5. The subgroups within the vegetable group in the USDA Food Guide include:

   a.
   b.
   c.
   d.
   e.

6. Food labels must provide the following information on the panel called *"Nutrition Facts."*

   a.   g.
   b.   h.
   c.   i.
   d.   j.
   e.   k.
   f.   l.

7. Identify the number of calories provided by 1 exchange in each of the following exchange lists.

    a. Vegetable   = _____ calories
    b. Starch   = _____ calories
    c. Nonfat milk   = _____ calories
    d. Fat   = _____ calories
    e. Fruit   = _____ calories
    f. Medium-fat meat = _____ calories
    g. Low-fat milk   = _____ calories
    h. High-fat meat   = _____ calories
    i. Whole milk   = _____ calories
    j. Lean meat   = _____ calories

8. List 5 ways someone may choose to "spend" their discretionary calorie allowance:

    a.
    b.
    c.
    d.
    e.

Solve these problems:

1. Use the menu below to complete sections a-d which follow:

Menu

Breakfast	Lunch	Dinner
1 hard boiled egg	2 tbs. peanut butter	3 oz. hamburger
1 cup cornflakes	1 tsp. jelly	hamburger bun
1 cup skim milk	2 slices bread	2 tbs. low-fat mayonnaise
1 medium grapefruit	1 medium apple	½ cup green beans
	1 diet cola	1 cup plain, nonfat yogurt

    a. Categorize the foods in the menu according to the USDA Food Guide:

Fruits	Vegetables	Grains	Meat, Poultry, Fish, Legumes, Eggs, and Nuts	Milk, Yogurt, and Cheese

    b. Which foods were you unable to classify?

c. Evaluate the adequacy of the menu according to the USDA Food Guide by comparing the actual amounts with the recommended amounts of the following for a sedentary woman 32 years old with a daily allowance of 1,800 calories (see Table 2-3 on page 39).

Food Group	Recommended Amount	Actual Amount
Fruits	1 ½ cups	_____
Vegetables	2 ½ cups	_____
Grains	6 oz.	_____
Meat, poultry, fish, legumes, eggs, and nuts	5 oz.	_____
Milk, yogurt, and cheese	3 cups	_____
Oils	5 tsp.	_____

d. Categorize the foods in the menu into exchanges by placing an X in the appropriate blank.

	Starch	*Meats*	*Vegetables*	*Fruits*	*Milks*	*Fats*	*Other*
egg	____	____	____	____	____	____	____
cornflakes	____	____	____	____	____	____	____
skim milk	____	____	____	____	____	____	____
jelly	____	____	____	____	____	____	____
grapefruit	____	____	____	____	____	____	____
peanut butter	____	____	____	____	____	____	____
bread	____	____	____	____	____	____	____
apple	____	____	____	____	____	____	____
diet cola	____	____	____	____	____	____	____
hamburger	____	____	____	____	____	____	____
bun	____	____	____	____	____	____	____
mayonnaise	____	____	____	____	____	____	____
green beans	____	____	____	____	____	____	____
plain yogurt	____	____	____	____	____	____	____

2. a. How many calories are in a day's meal which provides 260 grams carbohydrate, 30 grams protein, and 60 grams of fat? _____ calories

b. What percent of the calories are from carbohydrate? _____ %

c. What percent of the calories are from fat? _____ %

d. How does the percentage of calories from fat compare to the amount in the *Healthy People 2010 Objectives?*

Answer these controversy questions:

1. What advice would you give someone who looks to consume chocolate to reduce the risk of heart attack and stroke?

2. Why do food manufacturers often refine away the natural flavonoids in foods?

3. What are some foods common to many Asian diets that contain phytoestrogens?

4. Which types of cancer occur less frequently in people who consume five tomato-containing meals per week, and why?

5. What are probiotics?

Study Aids

Use Tables 2-3 and 2-4 on page 39 and Figure 2-4 on pages 35-36 to study the USDA Food Guide, including the food groups, examples of foods in each group, nutrient density of foods in each group, and amounts of food recommended for specific target groups.

Sample Test Items

Comprehension Level Items:

1. Recommendations for dietary nutrient intakes of healthy people in the United States and Canada are called:

    a. Recommended Nutrient Intakes.
    b. Daily Values.
    c. Adequate Intakes.
    d. Dietary Reference Intakes.
    e. Tolerable Upper Intake Levels.

2. Which of the following establishes population-wide average requirements that researchers and policy makers use in their work?

    a. Adequate Intakes
    b. Tolerable Upper Intake Limits
    c. Estimated Average Requirements
    d. Recommended Dietary Allowances

3. All of the following adjectives are descriptive of the Dietary Reference Intakes **except:**

    a. minimal.
    b. approximate.
    c. recommended.
    d. generous.

4. The value set for energy is set at the average of the population's estimated energy requirements in order to:

    a. compensate for lack of physical activity.
    b. discourage overconsumption of energy which would lead to obesity.
    c. encourage a decrease in fat consumption.
    d. ensure generous allowances of fat and carbohydrate.

5. The term *Daily Values* appears on:

    a. fresh meats.
    b. refined foods.
    c. food labels.
    d. fresh vegetables.

6. Which of the following could not be part of a person's discretionary calorie allowance?

    a. calories from added sugars, such as jams, colas and honey
    b. calories from nutrient-dense foods needed to meet recommended nutrient intakes
    c. calories from added fat absorbed by the batter on a fried chicken leg
    d. calories from the naturally occurring fat in whole milk, which is consumed instead of nonfat milk

7. The USDA Food Guide recommends intakes of certain amounts from each food group based on age alone.

    a. true
    b. false

8. In the food exchange lists, all of the food portions on one list have about the same number of calories and about the same amount of energy nutrients.

   a. true
   b. false

9. In the exchange lists, ½ cup of any cooked vegetable in the vegetable exchange list provides about ____ calories.

   a. 25
   b. 45
   c. 60
   d. 80

10. In the exchange lists, meats and substitutes are separated into _____ categories, based on fat content.

    a. two
    b. three
    c. four
    d. five

11. Corn and potatoes are listed in the _____ list of the exchange system.

    a. meats
    b. vegetables
    c. starch
    d. fat

12. The exchange system highlights the fact that most foods provide more than one energy nutrient.

    a. true
    b. false

13. Based on the USDA Food Guide, which of the following would be recommended as primary sources of fats in the diet?

    a. fat that occurs naturally in milk, butter, sour cream
    b. fat that occurs naturally in the skin of poultry, fat streaks in beef, cream cheese
    c. lard, margarine, pudding made from whole milk
    d. olive oil, peanut oil, oils that occur naturally in fatty fish

14. The 2005 *Dietary Guidelines* advise Americans to increase consumption of all of the following **except**:

    a. refined grains.
    b. vegetables.
    c. fruits.
    d. fat-free or low-fat milk or milk products.

15. The American College of Sports Medicine recommends exercise for a duration of at least 30 minutes total daily.

    a. true
    b. false

16. Which of the following is the major purpose of the Daily Values listed on food labels?

    a. prevent chronic diseases
    b. allow comparisons among foods
    c. discourage overconsumption of calories
    d. prevent nutrient toxicity

17. A food with *healthy* as part of its name cannot contain any nutrient or food constituent in an amount known to increase disease risk.

    a. true
    b. false

18. Which of the following nutrient information is **not** required on food labels?

    a. fat in grams per serving
    b. polyunsaturated fat in grams per serving
    c. saturated fat in grams per serving
    d. cholesterol in milligrams per serving

19. Which of the following FDA grades for health claims, relating the roles of nutrients and food constituents to disease states, represents significant scientific agreement?

    a. A
    b. B
    c. C
    d. D

20. Until recently, FDA held manufacturers to the highest standards of scientific evidence before allowing them to place health claims on food labels.

    a. true
    b. false

21. The absence of a Tolerable Upper Intake Level for a nutrient does not imply that it is safe to consume it in any amount.

    a. true
    b. false

22. The DRI Committee recommends a diet that contains _____% of its calories from fat.

    a. 10-35
    b. 20-35
    c. 45-65
    d. 50-70

23. The DRI are **not** designed for:

    a. disease prevention.
    b. health restoration.
    c. repletion of nutrients.
    d. a and b
    e. b and c

24. Which of the following meals represents the **least** nutrient-dense Mexican food choices?

    a. corn tortilla, pinto beans, tomatoes, guava
    b. flour tortilla, black beans, iceburg lettuce, pineapple
    c. fried tortilla shell, refried beans, cheddar cheese, avocado
    d. lean chicken breast, squash, tomatoes, corn, mango

25. The USDA Food Guide can assist vegetarians in their food choices.

    a. true
    b. false

Application Level Items:

26. Which of the following is **not** one of the criteria used for establishing USDA serving equivalents?

    a. a specific, precise amount
    b. an amount typically served in restaurants
    c. an amount of food that delivers a known quantity of key nutrients
    d. an amount of food used in previous food guides to describe a serving

27. Based on the exchange lists in Appendix D, how many calories are contained in the following meal pattern for a lunch meal: 1 lean meat, 1 starch, 1 vegetable, 1 fruit, and 1 fat?

    a. 235
    b. 256
    c. 265
    d. 300

28. An ingredient list states that a food product contains peas, water, carrots, sugar, and salt. Which of these ingredients is present in the smallest quantity?

    a. peas
    b. carrots
    c. sugar
    d. salt
    e. water

29. You see the following ingredient list on a cereal label: *wheat, sugar, raisins, brown sugar, syrup, and salt.* You therefore conclude that this product:

    a. is a good source of fiber.
    b. contains 100% whole wheat.
    c. is a nutritious food choice.
    d. contains close to half its weight in sugar.

30. The Nutrition Facts Panel on a box of cereal lists the following information for amounts per serving: **167 calories; 27 calories from fat; 3 g total fat.** What percentage of the calories are provided by fat?

    a. 3%
    b. 11%
    c. 16%
    d. 27%

31. Stan wants to apply the concept of the discretionary calorie allowance to his diet. If he wants to attend his niece's birthday party and eat cake and ice cream tomorrow evening, what should he do?

   a. Stan should eat low-calorie, nutrient-dense foods from each food group during the day tomorrow, so that he will have enough discretionary calories left to spend on cake and ice cream after his nutrient needs are met.
   b. Stan should fast during the day so that he will have plenty of discretionary calories to spend on cake and ice cream at the party.
   c. Stan should not eat cake and ice cream at the party; only nutrient-dense foods may be eaten, and cake and ice cream are energy-dense but not nutrient-dense.
   d. Stan should eat whatever he wants tomorrow but keep track of the calories; it doesn't matter where his calories come from, as long as he doesn't consume more calories than his body uses.

Answers

Summing Up

(1) Canada; (2) Dietary Reference Intakes; (3) Adequate Intakes; (4) individuals; (5) disease prevention; (6) Estimated Average Requirements; (7) average; (8) Acceptable Macronutrient Distribution Ranges; (9) differ; (10) exchange; (11) subgroups; (12) nutrient density; (13) vegetables; (14) discretionary calorie allowance; (15) calories; (16) calories; (17) fat; (18) protein; (19) energy; (20) common; (21) net; (22) descending; (23) Nutrition Facts panel; (24) protein; (25) calcium; (26) Food and Drug Administration

Chapter Glossary
Matching Exercise: (1) b; (2) h; (3) c; (4) a; (5) i; (6) d; (7) e; (8) g; (9) f

Exercises

Chapter Study Questions:

1. Because too much energy is bad for health and leads to obesity, while too little energy may cause undernutrition.

2. The four goals of the committee include setting intake recommendations for individuals, facilitating nutrition research and policy, establishing safety guidelines, and preventing chronic diseases.

3. For most people, meeting the diet ideals of the *Dietary Guidelines* requires choosing **more** vegetables (especially dark green vegetables, orange vegetables, and legumes), fruits, whole grains and fat-free or low-fat milk and milk products. It also requires choosing **less** refined grains, total fats (especially saturated fat, *trans* fat, and cholesterol), added sugars, and total calories.

4. The discretionary calorie allowance is the balance of calories remaining in a person's energy allowance after accounting for the number of calories needed to meet recommended nutrient intakes through consumption of nutrient-dense foods. Any calories in a food which are not necessary in order to obtain its nutrients, including added fats, sugars, and naturally occurring fats in higher-fat versions of foods (such as whole milk, which could be replaced by nonfat milk), are considered discretionary. To apply this concept, a person must plan their diet by choosing nutrient-dense foods first, ensuring adequacy; once nutrient needs are met, any additional calories left in their allowance may be consumed as the person wishes, or omitted (to lose weight).

5. The DRI values vary from group to group, whereas on a food label, one set of values must apply to everyone. The Daily Values reflect the needs of an "average person."

6. USDA serving equivalents are specific, precise, and reliable for delivering certain amounts of key nutrients in foods. In contrast, portion sizes served in restaurants are random, and tend to be much larger than USDA serving equivalents.

Short Answer Questions:

1. (a) carbohydrate; (b) fat; (c) saturated fat; (d) protein

2. (a) take supplements; (b) consume foods and beverages to which vitamins or minerals are added

3. (a) Recommended Dietary Allowances; (b) Adequate Intakes; (c) Tolerable Upper Intake Levels; (d) Estimated Average Requirements; (e) Acceptable Macronutrient Distribution Ranges

4. (a) fruits; (b) vegetables; (c) grains; (d) meat, poultry, fish, legumes, eggs and nuts; (e) milk, yogurt and cheese

5. (a) dark green vegetables; (b) orange and deep yellow vegetables; (c) legumes; (d) starchy vegetables; (e) other vegetables

6. (a) serving size; (b) servings per container; (c) total calories and calories from fat; (d) fat grams per serving including saturated fat and trans fat; (e) cholesterol; (f) sodium; (g) total carbohydrate including fiber, starch and sugars; (h) protein; (i) vitamin A; (j) vitamin C; (k) calcium; (l) iron

7. (a) 25; (b) 80; (c) 90; (d) 45; (e) 60; (f) 75; (g) 120; (h) 100; (i) 150; (j) 55

8. (a) extra portions of nutrient-dense foods; (b) naturally-occurring or added fats; (c) added sugars; (d) alcohol; (e) omitting them from the diet

Problem-Solving:

1. a.

Fruits	Vegetables	Grains	Meat, Poultry, Fish, Legumes, Eggs, and Nuts	Milk, Yogurt, and Cheese
grapefruit apple	green beans	cornflakes bread hamburger bun	egg peanut butter hamburger	skim milk yogurt

b. jelly; diet cola; mayonnaise

c. fruits – 1 cup; vegetables – ½ cup; grains – 5 oz.; meat, poultry, fish, legumes, eggs, and nuts – 6 oz.; milk, yogurt, and cheese – 2 cups; oils – 2 tsp.

d.

	Starch	Meats	Vegetables	Fruits	Milks	Fats	Other
egg		X					
cornflakes	X						
skim milk					X		
jelly							X
grapefruit				X			
peanut butter		X					
bread	X						
apple				X			
diet cola							X
hamburger		X					
bun	X						
mayonnaise						X	
green beans			X				
plain yogurt					X		

2. (a) 1700 calories; (b) 61 % calories from carbohydrate; (c) 32% calories from fat; (d) the percent calories from fat is within the recommended level of 20-35%

Controversy Questions:

1. It would be better to obtain flavonoids from nutrient-dense food sources such as fruits, vegetables or green or black tea because chocolate promotes weight gain and contributes fat and sugar to the diet. However, chocolate can be used as an occasional treat within the context of the dietary principles of moderation, variety, and balance.

2. Because they impart a bitter taste to foods and most consumers desire milder flavors.

3. Soybeans and their products including tofu, miso and soy drink.

4. Cancers of the esophagus, prostate and stomach; because tomatoes contain lycopene, which has antioxidant activity and may inhibit reproduction of cancer cells.

5. Probiotics are microorganisms believed to alter the native bacterial colonies or other conditions in the digestive tract in ways that may reduce diseases.

Sample Test Items

1. d (p. 28)	9. a (Appendix D)	17. a (p. 51)	25. a (p. 44)
2. c (p. 29)	10. c (Appendix D)	18. b (p. 49)	26. b (p. 44)
3. a (p. 29-33)	11. c (p. 45)	19. a (p. 51)	27. c (Appendix D)
4. b (p. 33)	12. a (p. 45)	20. a (p. 50)	28. d (p. 47)
5. c (p. 28, 33)	13. d (p. 36, 38)	21. a (p. 30)	29. d (p. 47-50)
6. b (p. 37-38)	14. a (p. 34)	22. b (p. 30)	30. c (p. 48)
7. b (p. 38)	15. a (p. 34)	23. e (p. 31)	31. a (p. 37-38)
8. a (p. 45)	16. b (p. 33)	24. c (p. 43)	

Chapter 3 - The Remarkable Body

Chapter Objectives

After completing this chapter, you should be able to:

1. Identify the basic needs of cells and describe how cells are organized into tissues, organs, and systems.

2. Describe the major functions of the cardiovascular, hormonal, nervous, digestive, and excretory systems.

3. Explain the mechanical and chemical digestive processes in order of their occurrence in the body.

4. Differentiate between the mechanical and chemical aspects of digestion.

5. Discuss the processes of absorption, transportation, and storage of nutrients.

6. Describe alcohol's effects on the body and whether the benefits of alcohol outweigh the risks (Controversy 3).

Key Concepts

✓ The body's cells need energy, oxygen, and nutrients, including water, to remain healthy and do their work. Genes direct the making of each cell's machinery, including enzymes. Genes and nutrients interact in ways that affect health. Specialized cells are grouped together to form tissues and organs; organs work together in body systems.

✓ Blood and lymph deliver nutrients to all the body's cells and carry waste materials away from them. Blood also delivers oxygen to cells. The cardiovascular system ensures that these fluids circulate properly among all organs.

✓ Glands secrete hormones that act as messengers to help regulate body processes.

✓ The nervous system joins the hormonal system to regulate body processes through communication among all the organs. Together, the hormonal and nervous systems respond to the need for food, govern the act of eating, regulate digestion, and call for the stress response.

✓ The immune system enables the body to resist diseases.

✓ The preference for sweet, salty, and fatty tastes seems to be inborn and can lead to overconsumption of foods that offer them.

✓ The digestive tract is a flexible, muscular tube that digests food and absorbs its nutrients and some nonnutrients. Ancillary digestive organs aid digestion.

✓ The digestive tract moves food through its various processing chambers by mechanical means. The mechanical actions include chewing, mixing by the stomach, adding fluid, and moving the tract's contents by peristalsis. After digestion and absorption, wastes are excreted.

✓ Chemical digestion begins in the mouth, where food is mixed with an enzyme in saliva that acts on carbohydrates. Digestion continues in the stomach, where stomach enzymes and acid break down protein. Digestion then continues in the small intestine; there the liver and gallbladder contribute bile that emulsifies fat, and the pancreas and small intestine donate enzymes that continue digestion so that absorption can occur. Bacteria in the colon break down certain fibers.

✓ The healthy digestive system is capable of adjusting to almost any diet and can handle any combination of foods with ease.

✓ The mechanical and chemical actions of the digestive tract break foods down to nutrients, and large nutrients to their smaller building blocks, with remarkable efficiency.

✓ The digestive system feeds the rest of the body and is itself sensitive to malnutrition. The folds and villi of the small intestine enlarge its surface area to facilitate nutrient absorption through countless cells to the blood and lymph. These transport systems then deliver the nutrients to all the body's cells.

✓ The digestive tract has many ways to communicate its needs. By taking the time to listen, you will obtain a complete understanding of the mechanics of the digestive tract and its signals.

✓ The kidneys adjust the blood's composition in response to the body's needs, disposing of everyday wastes and helping remove toxins. Nutrients, including water, and exercise help keep the kidneys healthy.

✓ The body's energy stores are of two principal kinds: glycogen in muscle and liver cells (in limited quantities) and fat in fat cells (in potentially large quantities). Other tissues store other nutrients.

✓ To achieve optimal function, the body's systems require nutrients from outside. These have to be supplied through a human being's conscious food choices.

Summing Up

The body is composed of trillions of (1)_____ and the nature of their work is determined by genes. Among the cells' most basic needs are (2)_____, oxygen, and nutrients to remain healthy and do their work. Specialized cells are grouped together to form tissues and (3) _____. Body (4)_____ supply the tissues continuously with energy, oxygen and nutrients. The body's main fluids are the blood and (5)_____. Blood travels within the arteries, (6)_____ and capillaries, as well as within the heart's chambers.

As the blood travels through the cardiovascular system, it picks up oxygen in the (7)_____ and also releases carbon dioxide there. As it passes through the digestive system, the blood picks up most (8)_____, other than fats, for distribution elsewhere. All blood leaving the digestive system is routed directly to the (9)_____. The blood is cleansed of its wastes as it passes through the (10)_____.

The hormonal and (11)_____ systems regulate body processes through communication among all the organs. They respond to the need for food, govern the act of eating, regulate (12)_____ and call for the stress response. The (13)_____ system's job is to digest food to its component nutrients and then to (14)_____ those nutrients, leaving behind the substances that are appropriate to excrete. To do this, the system works at two levels, one mechanical and the other (15)_____.

The body's cells need nutrients around the (16)_____ which requires systems of

(17)_____ and release to meet the cells' needs between meals. However, some

(18)_____ are stored in the body in much larger quantities than others. The liver converts excess

energy-containing nutrients into two forms, glycogen and (19)_____. Body stores for other

nutrients include the liver and fat cells which store many vitamins and the (20)_____ which

provide reserves of calcium and other minerals.

Chapter Glossary

Matching Exercise 1:

_____ 1. kidneys	a.	to break molecules into smaller molecules; a main function of the digestive tract with respect to food
_____ 2. liver	b.	the fluid of the cardiovascular system composed of water, red, and white blood cells, other formed particles, nutrients, oxygen, and other constituents
_____ 3. digest		
_____ 4. cortex	c.	the outermost layer of something
_____ 5. intestine	d.	blood vessels that carry blood, with the carbon dioxide it has collected, from the tissues back to the heart
_____ 6. norepinephrine	e.	a compound related to epinephrine that helps to elicit the stress response
_____ 7. veins		
_____ 8. digestive system	f.	a pair of organs that filter wastes from the blood, make urine, and release it to the bladder for excretion from the body
_____ 9. blood	g.	a large, lobed organ that lies just under the ribs
_____ 10. large intestine	h.	a long, tubular organ of digestion and the site of nutrient absorption
_____ 11. heartburn	i.	the body system composed of organs that break down complex food particles into smaller, absorbable products
_____ 12. pyloric valve	j.	the portion of the intestine that completes the absorption process
_____ 13. hernia	k.	the circular muscle of the lower stomach that regulates the flow of partly digested food into the small intestine
_____ 14. constipation	l.	a 20-foot length of small-diameter intestine that is the major site of digestion of food and absorption of nutrients
_____ 15. small intestine	m.	infrequent, difficult bowel movements often caused by diet, inactivity, dehydration, or medication
_____ 16. sphincter	n.	a protrusion of an organ or part of an organ through the wall of the body chamber that normally contains the organ
_____ 17. bladder	o.	a burning sensation in the chest (heart) area caused by backflow of stomach acid into the esophagus
	p.	a circular muscle surrounding, and able to close, a body opening
	q.	the sac that holds urine until time for elimination

Crossword Puzzle 1:

Crossword Puzzle 1 Word Bank:		
absorb	epinephrine	insulin
arteries	fat cells	organs
body system	genes	peristalsis
capillaries	glucagon	tissues
cells	hormones	
enzyme	hypothalamus	

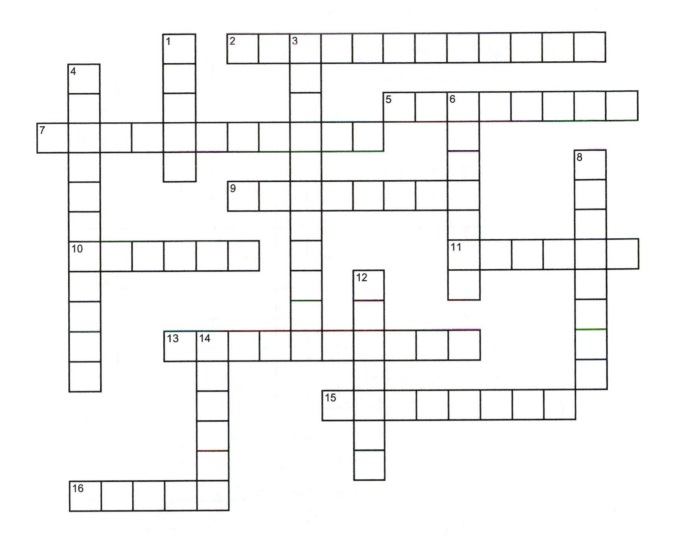

	Crossword Puzzle 1 Across:
2.	a part of the brain that senses a variety of conditions in the blood, such as temperature, glucose content, salt content, and others
5.	cells that specialize in the storage of fat and that form the fat tissue
7.	the major hormone that elicits the stress response
9.	blood vessels that carry blood containing fresh oxygen supplies from the heart to the tissues
10.	to take in, as nutrients are taken into the intestinal cells after digestion; the main function of the digestive tract with respect to nutrients
11.	a working protein that speeds up a specific chemical reaction
13.	a group of related organs that work together to perform a function
15.	a hormone from the pancreas that stimulates the liver to release glucose into the bloodstream
16.	the smallest units in which independent life can exist

	Crossword Puzzle 1 Down:
1.	units of a cell's inheritance, made of the chemical DNA (deoxyribonucleic acid)
3.	the wave-like muscular squeezing of the esophagus, stomach, and small intestine that pushes their contents along
4.	minute, weblike blood vessels that connect arteries to veins and permit transfer of materials between blood and tissues
6.	systems of cells working together to perform specialized tasks
8.	chemicals that are secreted by glands into the blood in response to conditions in the body that require regulation
12.	a hormone from the pancreas that helps glucose enter cells from the blood
14.	discrete structural units made of tissues that perform specific jobs, such as the heart, liver, and brain

Crossword Puzzle 2:

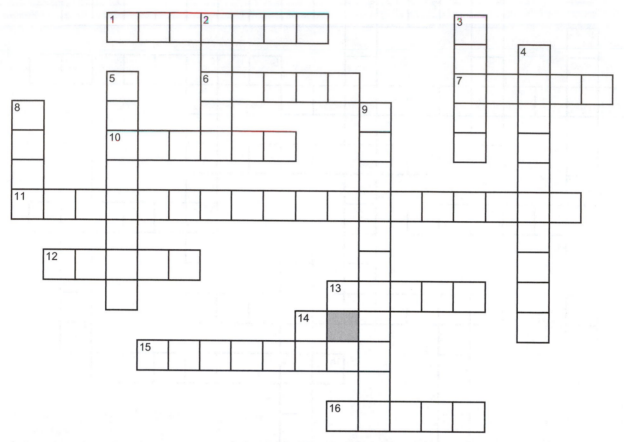

Crossword Puzzle 2 Word Bank:			
bicarbonate	diarrhea	lungs	pH
bile	emulsifier	lymph	plasma
chyme	extracellular fluid	mucus	stomach
colon	feces	nephrons	villi

Crossword Puzzle 2 Across:	Crossword Puzzle 2 Down:
1. muscular, elastic, pouchlike organ of the digestive tract that grinds and churns swallowed food and mixes it with acid and enzymes, forming chyme	2. a slippery coating of the digestive tract lining (and other body linings) that protect the cells from exposure to digestive juices (and other destructive agents)
6. the fluid resulting from the actions of the stomach upon a meal	3. fingerlike projections of the sheets of cells that line the intestinal tract
7. the fluid that moves from the bloodstream into tissue spaces and then travels in its own vessels, which eventually drain back into the bloodstream	4. a compound with both water-soluble and fat-soluble portions that can attract fats and oils into water, combining them
10. the cell-free fluid part of blood and lymph	5. the working units in the kidneys, consisting of intermeshed blood vessels and tubules
11. fluid residing outside the cells that transports materials to and from cells	8. a cholesterol-containing digestion fluid made by the liver, stored in the gallbladder, and released into the small intestine when needed
12. the body's organs of gas exchange	
13. the large intestine	9. a common alkaline chemical; a secretion of the pancreas
15. frequent, watery bowel movements usually caused by diet, stress, or irritation of the colon	
16. waste material remaining after digestion and absorption are complete	14. a measure of acidity on a point scale

Matching Exercise 2:

_____	1. antibodies	a.	lymphocytes that attack antigens
_____	2. metabolism	b.	a system of tissues and organs that defend the body against antigens, foreign materials that have penetrated the skin or body linings
_____	3. immune system	c.	the sum of all physical and chemical changes taking place in living cells
_____	4. microbes	d.	intermittent disturbance of bowel function, especially diarrhea or alternating diarrhea and constipation
_____	5. antigen	e.	chemicals that are released at the end of a nerve cell when a nerve impulse arrives
_____	6. lymphocytes	f.	proteins, made by cells of the immune system, that are expressly designed to combine with and to inactivate specific antigens
_____	7. pancreas	g.	bacteria, viruses, or other organisms invisible to the naked eye, some of which cause diseases
_____	8. neurotransmitters	h.	medications that react directly and immediately with the acid of the stomach, neutralizing it
_____	9. microvilli	i.	spasms of both the vocal cords and the diaphragm, causing periodic, audible, short, inhaled coughs
_____	10. phagocytes	j.	a microbe or substance that is foreign to the body
_____	11. hiccups	k.	white blood cells that can ingest and destroy antigens
_____	12. T-cells	l.	white blood cells that participate in the immune response; B-cells and T-cells
_____	13. irritable bowel syndrome	m.	lymphocytes that produce antibodies
_____	14. ulcer	n.	a severe and chronic splashing of stomach acid and enzymes into the esophagus, throat, mouth, or airway that causes inflammation and injury to those organs
_____	15. B-cells	o.	a storage form of carbohydrate energy
_____	16. gastro-esophageal reflux disease	p.	tiny, hairlike projections on each cell of every villus that can trap nutrient particles and absorb them into the cells
_____	17. antacids	q.	an organ with two main functions; one is an endocrine function and the other is an exocrine function
_____	18. glycogen	r.	an erosion in the topmost, and sometimes underlying, layers of cells that form a lining

Exercises

Answer these chapter study questions:

1. Identify and describe three factors necessary to ensure efficient circulation of fluid to all body cells.

2. Differentiate between the functions of glands and hormones.

3. Differentiate between the mechanical aspect of digestion and the chemical aspect of digestion.

4. Describe how the body's digestive system is affected by:

   a. severe undernutrition.

   b. lack of dietary fiber.

5. What is the difference between glycogen and fat?

6. Why do health care professionals inspect the skin and inside of the mouth to detect signs of malnutrition?

Complete these short answer questions:

1. The cells' most basic needs include:

   a.
   b.
   c.

2. The body's main fluids are the _____ and _____.

3. These are the central controllers of the nervous system:

   a.
   b.

4. The body's hormonal balance is altered by:

   a.
   b.
   c.

5. Five factors required to support the health of the kidneys include:

   a.
   b.
   c.
   d.
   e.

6. The liver converts excess energy containing nutrients into two forms including _____ and

   _____.

Solve these problems:

For questions 1-8, match the organs of the digestive system, listed on the left, to their primary functions, listed on the right.

_____ 1. stomach	a. manufactures enzymes to digest all energy-producing nutrients and releases bicarbonate to neutralize stomach acid
_____ 2. liver	b. stores bile until needed
_____ 3. large intestine	c. churns, mixes, and grinds food to a liquid mass
_____ 4. mouth	d. passes food to the stomach
_____ 5. pancreas	e. manufactures bile to help digest fats
_____ 6. esophagus	f. reabsorbs water and minerals
_____ 7. small intestine	g. secretes enzymes that digest carbohydrate, fat, and protein; absorbs nutrients
_____ 8. gallbladder	h. chews food and mixes it with saliva

Answer these controversy questions:

1. Why is the term moderation difficult to define in relationship to alcohol consumption?

2. How does food slow the absorption of alcohol?

3. What advice would you give someone who wants to drink alcohol socially but not become intoxicated?

4. Why are nutrient deficiencies an inevitable consequence of alcohol abuse?

5. Where else can the health-promoting components of red wine be found?

Study Aids

1. Use Table 3-1 on page 81 in your textbook to study the chemical digestion of nutrients.

2. On the following page, label the names of the accessory organs that aid digestion, on the left, and the names of the digestive tract organs that contain the food, on the right.

a. _____
Donate a starch-digesting
enzyme
Donate a trace of
fat-digesting enzyme
(important to infants)

b. _____
Manufactures bile, a
detergent-like substance
that facilitates digestion
of fats

c. _____
Stores bile until needed

d. _____
Conducts bile to
small intestine

e. _____
Conducts pancreatic
juice into small intestine

f. _____
Manufactures enzymes
to digest all energy-
yielding nutrients
Releases bicarbonate
to neutralize stomach
acid that enters
small intestine

g. _____
Chews and mixes food
with saliva

h. _____
Passes food to stomach

i. _____
Adds acid, enzymes,
and fluid
Churns, mixes, and grinds
food to a liquid mass

j. _____
Secretes enzymes that
digest carbohydrate, fat,
and protein
Cells lining intestine
absorb nutrients into blood
and lymph

k. _____
Reabsorbs water and
minerals
Passes waste (fiber,
bacteria, any unabsorbed
nutrients) and some
water to rectum

l. _____
Stores waste prior
to elimination

m. _____
Holds rectum closed
Opens to allow
elimination

Comprehension Level Items:

1. Genes direct the making of a piece of protein machinery which is most often a (an):

    a. antibody.
    b. hormone.
    c. enzyme.
    d. antigen.

2. Every cell continuously uses up:

    a. oxygen.
    b. nutrients.
    c. carbon dioxide.
    d. a and b
    e. b and c

3. All blood leaving the digestive system is routed directly to the:

    a. pancreas.
    b. liver.
    c. kidneys.
    d. gallbladder.

4. The blood often serves as an indicator of disorders caused by dietary deficiencies or imbalances of vitamins and minerals.

    a. true
    b. false

5. Hormones act as messengers that stimulate various organs to take appropriate actions.

    a. true
    b. false

6. When the pancreas detects a high concentration of the blood's sugar, glucose, it:

    a. secretes a hormone called glucagon.
    b. stimulates the liver to release glucose into the blood.
    c. stimulates special nerve cells in the hypothalamus.
    d. releases a hormone called insulin.

7. Which of the following statements is **not** true concerning the hormonal system?

    a. It regulates the menstrual cycle in women.
    b. It helps to regulate hunger and appetite.
    c. It is affected by nutrition very little.
    d. It regulates the body's reaction to stress.

8. Which of the following occur(s) as part of the stress response?

    a.   The pupils of the eyes widen.
    b.   The heart races to rush oxygen to the muscles.
    c.   The digestive system speeds up.
    d.   a and b
    e.   a and c

9. The mechanical job of digestion is to:

    a.   chew and crush foods.
    b.   add fluid.
    c.   secrete enzymes.
    d.   a and c
    e.   a and b

10. The major site for digestion of food and absorption of nutrients is the:

    a.   small intestine.
    b.   colon.
    c.   stomach.
    d.   liver.

11. Which of the following ensures that the digestive tract lining will **not** be digested?

    a.   enzymes
    b.   mucus
    c.   bile
    d.   hydrochloric acid

For questions 12-16, match the digestive organs, listed on the left, with their appropriate functions, listed on the right.

_____12. small intestine
_____13. stomach
_____14. pancreas
_____15. large intestine
_____16. anus

    a.   releases bicarbonate to neutralize stomach acid
    b.   churns, mixes, and grinds food to a liquid mass
    c.   cells of wall absorb nutrients into blood and lymph
    d.   passes any unabsorbed nutrients, waste (fiber, bacteria), and some water to rectum
    e.   manufactures bile to help digest fats
    f.   holds rectum closed

17. Which of the following is(are) the main task(s) of the colon?

    a.   to absorb fiber
    b.   to reabsorb water
    c.   to absorb some minerals
    d.   a and b
    e.   b and c

18. The digestive tract can adjust to whatever mixture of foods is presented to it.

    a.   true
    b.   false

19. Characteristics of the cells of the intestinal lining include:

   a. they are very efficient.
   b. they are selective.
   c. they are inefficient.
   d. a and b
   e. b and c

20. Which of the following nutrients should be consumed in intervals throughout the day based on the body's storage systems?

   a. fat
   b. carbohydrate
   c. sodium
   d. protein

21. Which of the following cells are the first to defend the body tissues against invaders?

   a. T-cells
   b. B-cells
   c. phagocytes
   d. lymphocytes

22. Most people have aversions to these tastes:

   a. salty.
   b. bitter.
   c. sour.
   d. a and b
   e. b and c

23. _____ digestion is the stomach's main function.

   a. Protein
   b. Fat
   c. Carbohydrate
   d. Mineral

24. Which of the following is/are considered the waste and water removal specialist(s) in the body?

   a. liver
   b. kidneys
   c. lungs
   d. bladder

25. Which of the following digestive organs does **not** release digestive juices?

   a. stomach
   b. pancreas
   c. liver
   d. small intestine
   e. large intestine

Application Level Items:

26. Why does all blood first circulate to the lungs and then return to the heart?

    a.   in order to provide all tissues with oxygenated blood fresh from the lungs
    b.   to deliver needed nutrients to the heart after disposing of carbon dioxide in the lungs
    c.   to receive impetus from heartbeats that push blood out to all other body tissues
    d.   a and b
    e.   a and c

27. To avoid constipation you would:

    a.   choose foods that provide enough fiber.
    b.   drink enough water.
    c.   change your diet quickly.
    d.   a and b
    e.   b and c

28. In a healthy individual, blood glucose levels are maintained by:

    a.   glands.
    b.   enzymes.
    c.   hormones.
    d.   a and b
    e.   a and c

29. Gurgling noises from the digestive tract are most likely the result of:

    a.   consuming too many high fiber foods.
    b.   gulping air with each swallow of food.
    c.   eating too much food at one time.
    d.   failing to relax after eating.

30. Which of the following would you recommend for someone complaining about heartburn?

    a.   Suggest that he/she take antacids to neutralize the acid.
    b.   Suggest that he/she drink milk to coat the lining of the stomach.
    c.   Suggest that he/she avoid irritating foods and lay down after meals.
    d.   Suggest that he/she eat smaller meals and drink liquids an hour before or after meals.

31. Your friend tells you that she has problems digesting fruit and meat when she combines them in her diet. What would be your response?

    a.   The digestive tract can't handle more than one task at a time.
    b.   Meat and fruit should never be eaten in combination.
    c.   After eating fruit, wait at least three hours before consuming meat.
    d.   The digestive system can adjust to whatever mixture of foods is presented to it.

Answers

(1) cells; (2) energy; (3) organs; (4) fluids; (5) lymph; (6) veins; (7) lungs; (8) nutrients; (9) liver; (10) kidneys; (11) nervous; (12) digestion; (13) digestive; (14) absorb; (15) chemical; (16) clock; (17) storage; (18) nutrients; (19) fat; (20) bones

Chapter Glossary

Matching Exercise 1: (1) f; (2) g; (3) a; (4) c; (5) h; (6) e; (7) d; (8) i; (9) b; (10) j; (11) o; (12) k; (13) n; (14) m; (15) l; (16) p; (17) q

Crossword Puzzle 1: (1) genes; (2) hypothalamus; (3) peristalsis; (4) capillaries; (5) fat cells; (6) tissues; (7) epinephrine; (8) hormones; (9) arteries; (10) absorb; (11) enzyme; (12) insulin; (13) body system; (14) organs; (15) glucagon; (16) cells

Crossword Puzzle 2: (1) stomach; (2) mucus; (3) villi; (4) emulsifier; (5) nephrons; (6) chyme; (7) lymph; (8) bile; (9) bicarbonate; (10) plasma; (11) extracellular fluid; (12) lungs; (13) colon; (14) pH; (15) diarrhea; (16) feces

Matching Exercise 2: (1) f; (2) c; (3) b; (4) g; (5) j; (6) l; (7) q; (8) e; (9) p; (10) k; (11) i; (12) a; (13) d; (14) r; (15) m; (16) n; (17) h; (18) o

Exercises

Chapter Study Questions:

1. Drinking sufficient water to replace water lost each day; maintaining cardiovascular fitness, which requires attention to nutrition and physical activity; and healthy red blood cells because they carry oxygen to all other cells, enabling them to use fuels for energy.

2. Glands are organs that monitor conditions that need regulation and produce hormones to regulate those conditions. Hormones are chemicals that are secreted by glands into the blood in response to conditions in the body that require regulation.

3. The mechanical aspects of digestion include chewing, mixing by the stomach, adding fluid, and moving the digestive tract's contents by peristalsis. The chemical aspects of digestion involve several organs of the digestive system that secrete special digestive juices that contain enzymes that break nutrients down into their component parts.

4. a. The absorptive surface of the small intestine shrinks and makes it impossible to obtain the few nutrients that the limited food supply may make available.

   b. The digestive tract muscles have too little undigested bulk to push against and get too little exercise, which makes them become weak.

5. Glycogen is stored in the liver and the muscles and it is a storage form of carbohydrate energy. Fat is a storage form of lipid energy and is stored in the fat cells. Fat tissue has virtually unlimited storage capacity, but the liver can only store a limited supply of glycogen.

6. Because vitamin and other nutrient deficiencies easily damage the skin, which serves as a physical barrier against infection, and the membranes, which line the body cavities and resist penetration by microbes.

Short Answer Questions:

1. (a) energy; (b) oxygen; (c) essential nutrients

2. blood; lymph

3. (a) brain; (b) spinal cord

4. (a) fasting; (b) feeding; (c) exercise

5. (a) a strong cardiovascular system; (b) abundant water supply; (c) sufficient energy; (d) exercise; (e) vitamins and minerals

6. glycogen; fat

Problem-Solving:

(1) c; (2) e; (3) f; (4) h; (5) a; (6) d; (7) g; (8) b

Controversy Questions:

1. Because no single upper limit of alcohol per day is appropriate for everyone and tolerances to alcohol differ.

2. Alcohol molecules in a stomach filled with food have a lower probability of touching the walls of the stomach and diffusing through; food also holds alcohol in the stomach longer, slowing its entry into the highly absorptive small intestine.

3. The person should eat high-carbohydrate snacks that slow alcohol absorption, and high-fat snacks that slow peristalsis to keep the alcohol in the stomach longer. The person should also add water or ice to drinks to dilute them and choose non-alcoholic beverages first and then every other round to quench thirst.

4. Because alcohol displaces food and interferes directly with the body's use of nutrients by disrupting every tissue's metabolism of nutrients.

5. Wine has a high potassium content that may lower high blood pressure; wine facilitates the absorption of potassium, calcium, phosphorous, magnesium and zinc; wine also contains flavonoids that are antioxidants that may protect the cardiovascular system against damage from oxidation that leads to heart disease. The potassium would also be found in red grapes and a variety of fruits, vegetables, and other food sources, as well as in dealcoholized wines. The flavonoids can also be found in onions and other vegetables.

Study Aids

2. (a) salivary glands; (b) liver; (c) gallbladder; (d) bile duct; (e) pancreatic duct; (f) pancreas; (g) mouth; (h) esophagus; (i) stomach; (j) small intestine; (k) large intestine; (l) rectum; (m) anus

Sample Test Items

1. c (p. 67)	12. c (p. 83)	23. a (p. 79)
2. d (p. 66)	13. b (p. 77-78)	24. b (p. 87)
3. b (p. 68)	14. a (p. 79)	25. e (p. 79)
4. a (p. 70)	15. d (p. 78-79)	26. e (p. 68-69)
5. a (p. 70-71)	16. f (p. 76)	27. d (p. 86-87)
6. d (p. 71)	17. e (p. 78)	28. e (p. 71)
7. c (p. 71-72)	18. a (p. 81)	29. b (p. 83, 85)
8. d (p. 73)	19. d (p. 83)	30. d (p. 85)
9. e (p. 77)	20. b (p. 88)	31. d (p. 81)
10. a (p. 79)	21. c (p. 73)	
11. b (p. 79)	22. e (p. 74)	

Chapter 4 – The Carbohydrates
Sugar, Starch, Glycogen, and Fiber

Chapter Objectives

After completing this chapter, you should be able to:

1. Distinguish among the various carbohydrates found in foods and in the human body.

2. Differentiate between complex and simple carbohydrates and their effects on the body.

3. Describe the body's use of glucose to provide energy or to make glycogen and fat.

4. Discuss diabetes, hypoglycemia, and lactose intolerance and their relationship to carbohydrate intake.

4. Discuss the roles of fiber-rich foods in the maintenance of the body's health and identify foods rich in fiber.

5. Assess the role sugar and alternative sweeteners play in one's diet (Controversy 4).

Key Concepts

✓ Through photosynthesis, plants combine carbon dioxide, water, and the sun's energy to form glucose. Carbohydrates are made of carbon, hydrogen, and oxygen held together by energy-containing bonds: *carbo* means "carbon"; *hydrate* means "water."

✓ Glucose is the most important monosaccharide in the human body. Most other monosaccharides and disaccharides become glucose in the body.

✓ Starch is the storage form of glucose in plants and is also nutritive for human beings.

✓ Glycogen is the storage form of glucose in animals and human beings.

✓ Human digestive enzymes cannot break the bonds in fiber, so most of it passes through the digestive tract unchanged. Some fiber, however, is susceptible to fermentation by bacteria in the colon.

✓ The body tissues use carbohydrates for energy and other functions; the brain and nerve tissues prefer carbohydrate as fuel. Nutrition authorities recommend a diet based on foods rich in complex carbohydrates and fiber.

✓ Fiber-rich diets benefit the body by helping to normalize blood cholesterol and blood glucose and by maintaining healthy bowel function. They are also associated with healthy body weight.

✓ Foods rich in soluble viscous fibers help control blood cholesterol.

✓ Foods rich in viscous fibers help to modulate blood glucose concentrations.

✓ Fibers in foods help to maintain digestive tract health.

✓ Diets that are adequate in fiber assist the eater in maintaining a healthy body weight.

✓ Most adults need between 21 and 38 grams of total fiber each day, but few consume this amount. Fiber needs are best met with whole foods. Purified fiber in large doses can have undesirable effects. Fluid intake should increase with fiber intake.

✓ With respect to starch and sugars, the main task of the various body systems is to convert them to glucose to fuel the cells' work. Fermentable fibers may release gas as they are broken down by bacteria in the intestine.

✓ Lactose intolerance is a common condition in which the body fails to produce sufficient amounts of the enzyme needed to digest the sugar of milk. Uncomfortable symptoms result and can lead to milk avoidance. Lactose-intolerant people and those allergic to milk need milk alternatives that contain the calcium and vitamins of milk.

✓ Without glucose, the body is forced to alter its uses of protein and fats. To help supply the brain with glucose, the body breaks down protein to make glucose and converts its fats into ketone bodies, incurring ketosis.

✓ Glycogen is the body's storage form of glucose. The liver stores glycogen for use by the whole body. Muscles have their own glycogen stock for their exclusive use. The hormone glucagon acts to liberate stored glucose from liver glycogen.

✓ Blood glucose regulation depends mainly on the hormones insulin and glucagon. Most people have no problem regulating their blood glucose when they consume mixed meals at regular intervals.

✓ The glycemic index is a measure of blood glucose response to foods relative to the response to a standard food. The glycemic load is the product of the glycemic index multiplied by the carbohydrate content of a food.

✓ The liver has the ability to convert glucose into fat; under normal conditions, most excess glucose is stored as glycogen or used to meet the body's immediate needs for fuel.

✓ Diabetes is an example of the body's abnormal handling of glucose. Inadequate or ineffective insulin leaves blood glucose high and cells undersupplied with glucose energy. Weight control and exercise may be effective in preventing the predominant form of diabetes (type 2) and the illnesses that accompany it. A person diagnosed with diabetes must establish patterns of eating, exercise, and medication to control blood glucose.

✓ Postprandial hypoglycemia is an uncommon medical condition in which blood glucose falls too low. It can be a warning of organ damage or disease. Many people believe they experience symptoms of hypoglycemia, but their symptoms do not accompany below-normal blood glucose.

Summing Up

Carbohydrates are the first link in the food chain and are obtained almost exclusively from

(1)_____. Through (2)_____, plants combine carbon dioxide, water, and the

sun's energy to form (3)_____, from which the human body can obtain energy. Carbohydrates are

made of carbon, hydrogen, and (4)_____. Six sugar molecules are important in nutrition, including three

(5)_____ and three disaccharides. The monosaccharides include glucose, galactose, and

(6)_____; the disaccharides include (7)_____, maltose, and

(8)_____.

Glucose may be strung together in long strands of thousands of glucose units to form

(9)_____, which include starch, glycogen, and most of the (10)_____. Some of the best known insoluble fibers are (11)_____ and hemicellulose, while soluble fibers include

(12)_____ and gums.

Glucose from carbohydrate is the preferred (13)_____ for most body functions; nerve cells,

including those of the (14)_____, depend almost exclusively on glucose for their energy.

Unfortunately, (15)_____ carbohydrate is often wrongly accused in popular books as being the

"fattening" ingredient of foods. The Committee on DRI recommends (16)_____ percent of total

calories from carbohydrates.

Many different forms of fiber exist, each with their specific effects on health. According to the American

Dietetic Association, most people need (17)_____ grams of total fiber each day. Fiber needs are

best met by eating (18)_____ plant foods, which contain a mix of fiber types;

(19)_____ fiber in large doses can have undesirable effects.

Glucose is the basic carbohydrate unit that each (20)_____ of the body uses for energy. If the

blood delivers more glucose than the cells need, the liver and (21)_____ cells take up the surplus and

build the polysaccharide (22)_____. Without sufficient carbohydrate, the body turns to

(23)_____ to make glucose. Also, without sufficient carbohydrate, the body cannot use its

(24)_____ in the normal way.

Some people have physical conditions which cause abnormal handling of carbohydrates. These conditions

include (25)_____ intolerance, hypoglycemia, and (26)_____. The predominant

type of diabetes is (27)_____, characterized by (28)_____ resistance of the body's

cells. In contrast, the person with type 1 diabetes secretes no (29)_____. The effects of diabetes

can be severe because the disease causes destruction or blockage of (30)_____ that feed the body

organs, and tissues die from lack of nourishment. Nutrition intervention plays a central role in managing diabetes,

with a focus in type 2 diabetes on (31)_____.

The term (32)_____ refers to abnormally low blood glucose. The two types of hypoglycemia

include postprandial and (33)_____. People who experience the symptoms of hypoglycemia may

benefit from eating regularly timed, balanced (34)_____-containing meals and minimizing

(35)_____ beverages.

Chapter Glossary

Matching Exercise 1:

_____ 1. monosaccharides

_____ 2. maltose

_____ 3. chelating agents

_____ 4. hemorrhoids

_____ 5. starch

_____ 6. appendicitis

_____ 7. polysaccharides

_____ 8. glucose

_____ 9. soluble fibers

_____ 10. diverticula

_____ 11. glycogen

_____ 12. fibers

_____ 13. sucrose

_____ 14. granules

_____ 15. insoluble fibers

a. compounds composed of long strands of glucose units linked together

b. molecules that attract or bind with other molecules and are therefore useful in either preventing or promoting movement of substances from place to place

c. a disaccharide composed of glucose and fructose

d. food components that readily dissolve in water and often impart gummy or gel-like characteristics to foods

e. a disaccharide composed of two glucose units

f. the tough, fibrous structures of fruits, vegetables, and grains; indigestible food components that do not dissolve in water

g. a plant polysaccharide composed of glucose; highly digestible by human beings

h. a polysaccharide composed of glucose, made and stored by liver and muscle tissues of human beings and animals as a storage form of glucose

i. packages of starch molecules; various plant species make granules of varying shapes

j. inflammation and/or infection of the appendix, a sac protruding from the intestine

k. a single sugar used in both plant and animal tissues for energy

l. sacs or pouches that balloon out of the intestinal wall, caused by weakening of the muscle layers that encase the intestine

m. the indigestible parts of plant foods, comprised mostly of cellulose, hemicellulose, and pectin

n. swollen, hardened (varicose) veins in the rectum, usually caused by the pressure resulting from constipation

o. single sugar units

Matching Exercise 2:

_____ 1. constipation

_____ 2. disaccharides

_____ 3. galactose

_____ 4. lactase

_____ 5. fructose

_____ 6. impaired glucose tolerance

_____ 7. sugars

_____ 8. carbohydrates

_____ 9. lactose

_____ 10. glycemic index

_____ 11. type 2 diabetes

_____ 12. insulin

_____ 13. ketone bodies

_____ 14. ketosis

_____ 15. resistant starch

a. blood glucose levels higher than normal but not high enough to be diagnosed as diabetes

b. a disaccharide composed of glucose and galactose; sometimes known as milk sugar

c. pairs of single sugars linked together

d. a monosaccharide; sometimes known as fruit sugar

e. monosaccharide; part of the disaccharide lactose

f. difficult, incomplete, or infrequent bowel movements; associated with discomfort in passing feces from the body

g. an undesirable high concentration of ketone bodies, such as acetone, in the blood or urine

h. acidic, fat-related compounds that can arise from the incomplete breakdown of fat when carbohydrate is not available

i. hormone secreted by the pancreas in response to a high blood glucose concentration

j. the intestinal enzyme that splits the disaccharide lactose to monosaccharides during digestion

k. the type of diabetes in which the person makes plenty of insulin but the body cells resist insulin's action

l. simple carbohydrates; that is, molecules of either single sugar units or pairs of those sugar units bonded together

m. ranking of foods according to their potential for raising blood glucose relative to a standard such as glucose or white bread

n. compounds composed of single or multiple sugars

o. the fraction of starch in a food that is digested slowly, or not at all, by human enzymes

Crossword Puzzle:

Across:	Down:
9. hypoglycemia that occurs after 8 to 14 hours of fasting	1. an unusual drop in blood glucose that follows a meal and is accompanied by symptoms of the stress response
10. a disease characterized by elevated blood glucose and inadequate or ineffective insulin, which impairs a person's ability to regulate blood glucose normally	2. the action of carbohydrate and fat in providing energy that allows protein to be used for purposes it alone can serve
11. a hormone secreted by the pancreas that stimulates the liver to release glucose into the blood when blood glucose concentration dips	3. long chains of sugar units arranged to form starch or fiber; also called _polysaccharides_
13. the process by which green plants make carbohydrates from carbon dioxide and water using the green pigment chlorophyll to capture the sun's energy	4. a condition in which a normal amount of insulin produces a less than normal response by the tissues; metabolic consequence of obesity
14. the type of diabetes in which the pancreas produces no or very little insulin	5. the green pigment of plants that captures energy from sunlight for use in photosynthesis
	6. impaired ability to digest lactose due to a lack of the enzyme lactase
	7. sugars, including both single sugar units and linked pairs of sugar units
	8. a blood glucose concentration below normal, a symptom that may indicate any of several diseases, including impending diabetes
	12. a small fat fragment produced by the fermentary action of bacteria on viscous, soluble fibers

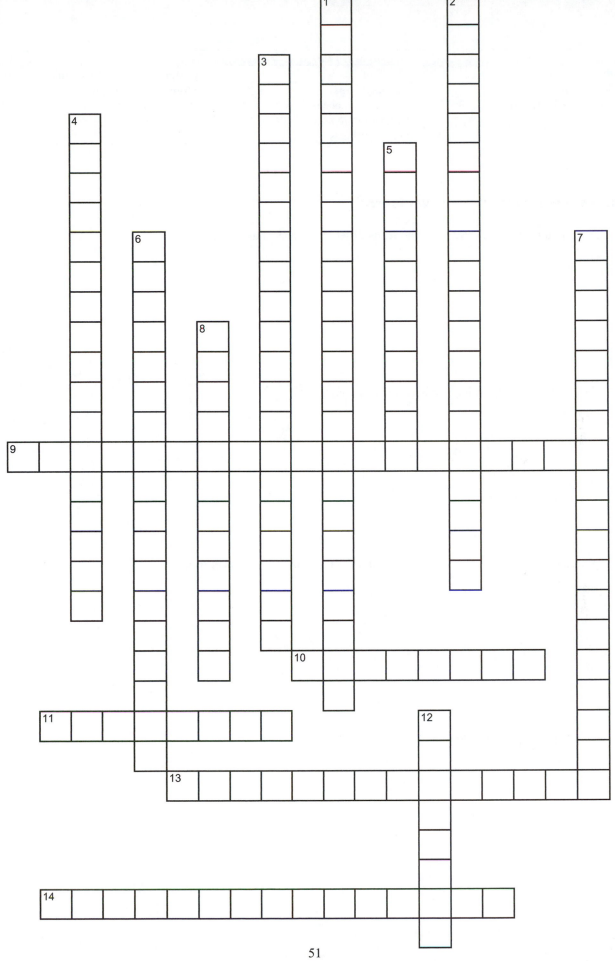

Exercises

Answer these chapter study questions:

1. From a nutrition perspective, how do fruits differ from concentrated sweets?

2. How does fiber help maintain a healthy body weight?

3. Describe how the body adjusts after a meal when the blood glucose level rises.

4. What happens when the body faces a severe carbohydrate deficit?

5. Why can some people with lactose intolerance consume and tolerate yogurt or aged cheese?

6. Describe insulin resistance that is a characteristic of type 2 diabetes.

Complete these short answer questions:

1. The only animal-derived food that contains significant amounts of carbohydrate is _____.

2. Carbohydrates are made of:

    a.
    b.
    c.

3. The monosaccharides include:

    a.
    b.
    c.

4. The disaccharides include:

    a.
    b.
    c.

5. The polysaccharides include:

    a.
    b.
    c.

6. The fibers of a plant contribute the supporting structures of its:

   a.
   b.
   c.

7. Soluble fiber is commonly found in:

   a.   d.
   b.   e.
   c.

8. Groups of individuals who are especially vulnerable to the adverse effects of excess fiber include:

   a.
   b.
   c.

9. A balanced diet that is high in _____ carbohydrate helps control body weight and maintain lean tissue.

10. An average-size person needs around _____ grams of carbohydrate per day to insure complete sparing of body protein and avoidance of ketosis.

Solve these problems:

1. Estimate the grams of dietary fiber and carbohydrate supplied by the following menu. Refer to the fiber information on page 126 for grams of fiber and the exchange information in Appendix D for grams of carbohydrate.

Breakfast	*Fiber (grams)*	*Carbohydrate (grams)*
1 cup skim milk		
1 slice whole wheat toast	_____	_____
1 oz. cheddar cheese	_____	_____
1 medium orange	_____	_____
	_____	_____
Lunch		
Salad with:		
2 cups raw lettuce		
1 medium tomato	_____	_____
2 tbsp. oil and vinegar dressing	_____	_____
1 medium apple	_____	_____
	_____	_____
Dinner		
3 oz. chicken breast		
½ cup broccoli	_____	_____
1 medium baked potato without skin	_____	_____
1 slice light rye bread	_____	_____
1 cup plain yogurt with	_____	_____
½ cup blackberries	_____	_____
	_____	_____
Total:	_____	_____

54

2. Use the ingredient label below to respond to the questions which follow.

INGREDIENTS: Bleached flour, sugar, partially hydrogenated vegetable shortening, dextrose, water, corn syrup, carob, whey blend, cornstarch, salt, sodium bicarbonate, lecithin, artificial flavoring and artificial colors.

a. List all of the sugars which appear in the above product.

b. Would you consider this product to be high or low in sugar content? Why?

Answer these controversy questions:

1. What is primarily responsible for the steady, upward trend in U.S. sugar consumption?

2. Why does fat and total calorie intake usually rise when sugar intake increases?

3. How would you respond to someone who makes the statement that sugary foods cause children to become hyperactive and unruly?

4. If a person consumes 3,000 calories a day, how many calories from added sugars could he/she consume to be in compliance with the World Health Organization's recommendations?

5. Why are sugar alcohols and artificial sweeteners safer for teeth than carbohydrate sweeteners?

Study Aids

1. Identify whether the following health effects are characteristic of insoluble or soluble fiber by placing an X in the appropriate column.

	Insoluble	Soluble
a. slows transit of food through the upper digestive tract	_____	_____
b. alleviates constipation	_____	_____
c. slows glucose absorption	_____	_____
d. lowers blood cholesterol	_____	_____
e. lowers risk of diverticulosis	_____	_____

2. Use Table 4-2 on page 106 to study the characteristics, food sources, and health effects of soluble and insoluble fiber. Study Table 4-8 on page 122 in your textbook to differentiate between type 1 and type 2 diabetes.

Sample Test Items

Comprehension Level Items:

1. Carbohydrate-rich foods are obtained almost exclusively from animals.

   a. true
   b. false

2. Carbohydrates are made of the following:

   a. oxygen, nitrogen, and carbon.
   b. nitrogen, hydrogen, and carbon.
   c. hydrogen, carbon, and oxygen.
   d. carbon dioxide, hydrogen, and oxygen.

3. Fructose occurs in all of the following **except**:

   a. honey.
   b. table sugar.
   c. fruit.
   d. milk.

4. Maltose consists of:

   a. glucose and galactose.
   b. galactose and fructose.
   c. two glucose units.
   d. lactose and glucose.

5. Which of the following rarely occurs free in nature?

   a. fructose
   b. galactose
   c. sucrose
   d. glucose

6. Complex carbohydrates are called:

   a. natural sugars.
   b. monosaccharides.
   c. polysaccharides.
   d. complex sugars.

7. The storage form of glucose found in plants is called:

   a. starch.
   b. chlorophyll.
   c. fiber.
   d. glycogen.

8. Common uses of pectin include:

    a. to thicken jelly.
    b. to add a pleasing consistency to foods.
    c. to make milk more digestible.
    d. a and b
    e. a and c

For questions 9–11, match the terms on the right with their definitions on the left.

9. _____ food components that readily dissolve in water and often impart gummy or gel-like characteristics to foods

10. _____ the indigestible parts of plant foods, largely non-starch polysaccharides

11. _____ the tough, fibrous structures of fruits, vegetables, and grains

a. fibers
b. residue
c. roughage
d. insoluble fibers
e. soluble fibers

12. The human brain depends almost exclusively on _____ for its energy.

    a. fat
    b. carbohydrate
    c. alcohol
    d. protein

13. Fiber promotes maintenance of proper weight through all of the following effects **except:**

    a. absorbing water and creating a feeling of fullness.
    b. donating few calories.
    c. speeding up movement of food through the upper digestive tract.
    d. displacing calorie-dense concentrated fats and sweets from the diet.

14. Which of the following parts of a wheat kernel is rich in both nutrients and fiber?

    a. endosperm
    b. germ
    c. husk
    d. bran

15. The body's form of stored glucose is called:

    a. glycogen.
    b. starch.
    c. glucagon.
    d. ketones.

16. Which of the following would you choose to lower your blood cholesterol level?

    a. purified fiber
    b. wheat bran
    c. brown rice
    d. oat bran

17. Which of the following foods would be most easily digested based on its starch content?

   a. corn
   b. lima beans
   c. white bread
   d. potato

18. On drinking milk, a person with lactose intolerance is likely to experience:

   a. vomiting.
   b. diarrhea.
   c. excessive gas.
   d. a and b
   e. b and c

19. Possible effects of diabetes include:

   a. disease of the feet and legs.
   b. cataracts in the eye.
   c. cancer of the pancreas.
   d. a and b
   e. b and c

20. Carbohydrate-containing foods appear in all of the following food groups **except:**

   a. milk, cheese and yogurt.
   b. vegetables.
   c. grains.
   d. fats and oils.

21. Any one of the following fruit choices contributes about _____ grams of carbohydrate: ½ cup juice, a small banana, or ½ cup canned fruit.

   a. 5
   b. 12
   c. 15
   d. 18

22. Neither honey nor sugar is a good source of any nutrients other than carbohydrates.

   a. true
   b. false

23. The glycemic index may be important to people with:

   a. diabetes.
   b. hypoglycemia.
   c. lactose intolerance.
   d. diverticulosis.

24. The most appropriate diet for managing diabetes is:

   a. controlled in total carbohydrate.
   b. low in saturated fat.
   c. high in protein.
   d. a and b
   e. a and c

25. To magnify the sweetness of foods without adding extra sugar and boosting the calories, you would:

   a. serve sweet food cold.
   b. add sweet spices such as cinnamon and nutmeg.
   c. add a tiny pinch of salt.
   d. a and b
   e. b and c

Application Level Items:

26. Someone makes the statement that eating fruit is the same as eating a candy bar. Your response would be:

   a. fruits differ from candy bars in nutrient density.
   b. the sugars of fruits are diluted in large volumes of water.
   c. candy bars are packaged in fiber.
   d. a and b
   e. b and c

27. Which of the following would you recommend as a calcium source for someone with lactose intolerance?

   a. milk
   b. yogurt with added milk solids
   c. aged cheese
   d. a and b
   e. b and c

28. Why would you recommend that someone **not** add purified fiber to the diet?

   a. It might displace nutrients from the diet.
   b. It might easily be taken to the extreme.
   c. It is very expensive and adversely affects the taste of foods.
   d. a and b
   e. b and c

29. Which of the following types of flour contains all of the nutritive portions of the original grain?

   a. white flour
   b. unbleached flour
   c. whole grain flour
   d. wheat flour

30. What argument(s) would you use against following a low-carbohydrate diet?

    a. The body uses protein to make energy.
    b. The body uses up its fat as an energy source.
    c. The body cannot use its fat in the normal way.
    d. a and b
    e. a and c

31. Use the U.S. Food Exchange System in Appendix D of your textbook to determine the grams of carbohydrate and calories from carbohydrate in the following meal: **1 cup nonfat skim milk; 3 oz. sirloin steak;1/3 cup white rice; ½ cup cooked carrots; 1 slice whole wheat bread; 1 small apple.**

    a. 52 grams CHO; 208 calories
    b. 58 grams CHO; 232 calories
    c. 62 grams CHO; 248 calories
    d. 72 grams CHO; 288 calories

32. Based on the Daily Value, approximately 60% of total calories should come from carbohydrate. For someone consuming 2500 calories a day, this would equal _____ grams of carbohydrate.

    a. 250
    b. 375
    c. 400
    d. 525

Answers

Summing Up

(1) plants; (2) photosynthesis; (3) glucose; (4) oxygen; (5) monosaccharides; (6) fructose; (7) sucrose; (8) lactose; (9) polysaccharides; (10) fibers; (11) cellulose; (12) pectin; (13) fuel; (14) brain; (15) complex; (16) 45-65; (17) 20-35; (18) unrefined; (19) purified; (20) cell; (21) muscle; (22) glycogen; (23) protein; (24) fat; (25) lactose; (26) diabetes; (27) type 2;(28) insulin; (29) insulin; (30) capillaries; (31) weight control; (32) hypoglycemia; (33) fasting; (34) protein; (35) alcoholic

Chapter Glossary

Matching Exercise 1: (1) o; (2) e; (3) b; (4) n; (5) g; (6) j; (7) a; (8) k; (9) d; (10) l; (11) h; (12) m; (13) c; (14) i; (15) f
Matching Exercise 2: (1) f; *(2)* c; (3) e; (4) j; (5) d; (6) a; (7) l; (8) n; (9) b; (10) m; (11) k; (12) i; (13) h; (14) g; (15) o
Crossword Puzzle: (1) postprandial hypoglycemia; (2) protein-sparing action; (3) complex carbohydrates; (4) insulin resistance; (5) chlorophyll; (6) lactose intolerance; (7) simple carbohydrates; (8) hypoglycemia; (9) fasting hypoglycemia; (10) diabetes; (11) glucagon; (12) butyrate; (13) photosynthesis; (14) type one diabetes

Exercises

Chapter Study Questions:

1. The sugars in fruits are diluted in large volumes of water and packaged with fiber, phytochemicals, vitamins and minerals needed by the body. Concentrated sweets are much less nutrient-dense and may also be high in fat.

2. Because high-fiber foods absorb water from the digestive juices, swell and create a feeling of fullness; in addition, they donate few calories and displace calorie-dense concentrated fats and sweets from the diet. Some

fibers also slow movement of food through the upper digestive tract so that you feel fuller for a longer period of time.

3. The pancreas detects the excess glucose and releases the hormone insulin, which signals the body's tissues to take up surplus glucose. The body siphons off the excess into the liver, to be converted to glycogen or fat and into the muscle, to be converted to glycogen.

4. When there is a severe carbohydrate deficit, the body turns to protein to make glucose, thus diverting protein from its vitally important functions such as maintaining the body's immune defenses. Secondly, the body cannot use its fat in the normal way without sufficient carbohydrate and goes into ketosis, which disturbs the body's normal acid-base balance.

5. Because the bacteria or molds that help create yogurt and aged cheese digest lactose as they convert milk to the fermented product.

6. Insulin resistance of the body's cells means that insulin may be present in varying amounts, but it is not effective in moving glucose from the bloodstream into the cells.

Short Answer Questions:

1. milk

2. (a) carbon; (b) hydrogen; (c) oxygen

3. (a) glucose; (b) galactose; (c) fructose

4. (a) sucrose; (b) maltose; (c) lactose

5. (a) starch, (b) fiber; (c) glycogen

6. (a) leaves; (b) stems; (c) seeds

7. (a) barley; (b) legumes; (c) fruits; (d) oats; (e) vegetables

8. (a) malnourished; (b) elderly; (c) children who consume no animal products

9. complex

10. 130

Problem-Solving:

1.

Breakfast	*Fiber (grams)*	*Carbohydrate (grams)*
1 cup skim milk	___	12
1 slice whole wheat toast	2	15
1oz. cheddar cheese	___	___
1 medium orange	3	15

Lunch		
Salad with:		
2 cups raw lettuce	2	10
1 medium tomato	1	5
2 tbsp. oil and vinegar dressing	___	___
1 medium apple	4	15

Dinner

3 oz. chicken breast	___	___
½ cup broccoli	2	5
1 medium baked potato without skin	2	15
1 slice light rye bread	2	15
1 cup plain yogurt with	___	12
½ cup blackberries	4	15
Total:	*22*	*134*

2. (a) sugar; dextrose; corn syrup; (b) high; because it contains three different sugars which appear as the second, fourth, and sixth ingredients on the label

Controversy Questions:

1. A dramatic increase in the purchase of commercially prepared foods and beverages to which sugars have been added, including soft drinks and sugar-sweetened fruit drinks.

2. Because sweet treats often contain abundant fat and calories and people with diets high in added sugars consume more calories each day than people with lower sugar intakes.

3. Studies have overwhelmingly failed to demonstrate any consistent effects of sucrose on behavior in either normal or hyperactive children.

4. 300 calories from added sugar

5. Because bacteria in the mouth cannot metabolize sugar alcohols and artificial sweeteners as rapidly as sugar, so they do not contribute to dental caries.

Study Aids

		Insoluble	Soluble
1.	a. slows transit of food through the upper digestive tract	___	X
	b. alleviates constipation	X	___
	c. slows glucose absorption	___	X
	d. lowers blood cholesterol	___	X
	e. lowers risk of diverticulosis	X	___

Sample Test Items

1. b (p. 100)	9. e (p. 103)	17. c (p. 113-114)	25. e (p. 129)
2. c (p. 100)	10. a (p. 102)	18. e (p. 116)	26. d (p. 105, 125-129)
3. d (p. 101)	11. d (p. 103)	19. d (p. 123)	27. c (p. 116)
4. c (p. 101)	12. b (p. 104)	20. d (p. 125-127)	28. d (p. 112)
5. b (p. 101)	13. c (p. 104, 109)	21. c (p. 125)	29. c (p. 110)
6. c (p. 102)	14. d (p. 110)	22. a (p. 127-129)	30. e (p. 117-118)
7. a (p. 102)	15. a (p. 118)	23. a (p. 119)	31. c (p. Appendix D)
8. d (p. 103)	16. d (p. 106)	24. d (p. 123)	32. b (p. 104-105)

Chapter 5 – The Lipids
Fats, Oils, Phospholipids, and Sterols

Chapter Objectives

After completing this chapter, you should be able to:

1. Explain the ways lipids are useful, both in foods and in the body.

2. Describe the structure of a triglyceride noting the differences between saturated and unsaturated fats.

3. Summarize the processes of lipid digestion, absorption, transport, and utilization in the body, including the significance of the lipoproteins.

4. Explain why manufacturers frequently hydrogenate fats and the possible health implications of consuming the *trans* fatty acids formed during hydrogenation.

5. List the arguments for and against the use of fat replacers.

6. Plan a diet containing fat in the right kinds and in the recommended amounts to provide optimal health and pleasure in eating.

7. Describe the most recent dietary guidelines for fat and characteristics of the Mediterranean diet (Controversy 5).

Key Concepts

✓ Lipids not only serve as energy reserves but also cushion the vital organs, protect the body from temperature extremes, carry the fat-soluble nutrients and phytochemicals, serve as raw materials, and provide the major component of which cell membranes are made.

✓ Lipids provide more energy per gram than carbohydrate and protein, enhance the aromas and flavors of foods, and contribute to satiety, or a feeling of fullness, after a meal.

✓ The body combines three fatty acids with one glycerol to make a triglyceride, its storage form of fat. Fatty acids in food influence the composition of fats in the body.

✓ Fatty acids are energy-rich carbon chains that can be saturated (filled with hydrogens) or monounsaturated (with one point of unsaturation) or polyunsaturated (with more than one point of unsaturation). The degree of saturation of the fatty acids in a fat determines the fat's softness or hardness.

✓ Phospholipids, including lecithin, play key roles in cell membranes; sterols play roles as part of bile, vitamin D, the sex hormones, and other important compounds.

✓ In the stomach, fats separate from other food components. In the small intestine, bile emulsifies the fats, enzymes digest them, and the intestinal cells absorb them.

✓ Small lipids travel in the bloodstream unassisted. Large lipids are incorporated into chylomicrons for transport in the lymph and blood. Blood and other body fluids are watery, so fats need special transport vehicles—the lipoproteins—to carry them in these fluids.

✓ When low on fuel, the body draws on its stored fat for energy. Carbohydrate is necessary for the complete breakdown of fat.

✓ The chief lipoproteins are chylomicrons, VLDL, LDL, and HDL. Blood LDL and HDL concentrations are among the major risk factors for heart disease.

✓ Elevated blood cholesterol is a risk factor for cardiovascular disease. Among major dietary factors that raise blood cholesterol, saturated fat and *trans* fat intakes are the most influential. Dietary cholesterol raises blood cholesterol to a lesser degree.

✓ Trimming fat from food trims calories and, often, saturated and *trans* fat as well.

✓ Dietary measures to lower LDL in the blood involve reducing saturated fat and *trans* fat and substituting monounsaturated and polyunsaturated fats. Cholesterol-containing foods are nutritious and are best used in moderation by most people.

✓ Two polyunsaturated fatty acids, linoleic acid (an omega-6 acid) and linolenic acid (an omega-3 acid), are essential nutrients used to make substances that perform many important functions. The omega-6 family includes linoleic acid and arachidonic acid. The omega-3 family includes linolenic acid, EPA, and DHA. The principal food source of EPA and DHA is fish, but some species have become contaminated with environmental pollutants.

✓ Vegetable oils become more saturated when they are hydrogenated. Hydrogenated fats resist rancidity better, are firmer textured, and have a higher smoking point than unsaturated oils; but they also lose the health benefits of unsaturated oils.

✓ The process of hydrogenation also creates *trans* fatty acids. *Trans* fats act somewhat like saturated fats in the body.

✓ Fats added to foods during preparation or at the table are a major source of fat in the diet.

✓ Meats account for a large proportion of the hidden fat in many people's diets. Most people consume meat in larger amounts than recommended.

✓ The choice between whole and fat-free milk products can make a large difference to the fat and saturated fat content of a diet. Cheeses are a major contributor of saturated fat.

✓ Fat in breads and cereals can be well hidden. Consumers must learn which foods of this group contain fats.

Summing Up

The lipids in foods and in the human body fall into (1)_____ classes. About 95 percent of the

lipids are (2)_____. Other classes of the lipid family include the phospholipids and the

(3)_____.

Although too much fat in the diet has the potential to be (4)_____, lipids serve many valuable

functions in the body. Fat is the body's chief storage form for (5)_____ from food eaten in excess

of need. In addition, fat surrounds and cushions all the body's vital (6)_____ and insulates the body from extremes in (7)_____. Some essential nutrients, such as the essential (8)_____ and the fat-soluble vitamins A, D, E, and (9)_____, are found primarily in foods that contain fat.

Fatty acids are the major constituent of (10)_____, which are the chief form of fat, and they differ from one another in two ways: in (11)_____ length and in degree of (12)_____. The more unsaturated a fat, the more (13)_____ it is at room temperature. Generally, vegetable and (14)_____ oils are rich in polyunsaturates. However, two vegetable oils, including coconut and (15)_____, are highly saturated.

Within the body, many fats travel from place to place in blood as passengers in (16)_____. A lipoprotein's density changes with its lipid and (17)_____ contents. Two types of lipoproteins of concern to health care providers include (18)_____ and (19)_____. Elevated HDL concentrations are associated with a (20)_____ risk of heart attack. High blood (21)_____ is an indicator of risk for CVD and the main dietary factors associated with elevated blood cholesterol are high (22)_____ fat and *trans* fat intakes.

Two fatty acids, including linoleic and (23)_____, must be supplied by the diet and are, therefore, (24)_____ fatty acids. Points of unsaturation in fatty acids are vulnerable to attack by (25)_____ and when the unsaturated points are oxidized, the oils become (26)_____. Cooking oils should be stored in tightly covered containers that exclude (27)_____. One way to prevent spoilage of unsaturated fats and also to make them harder is to change their fatty acids chemically by (28)_____. The process of hydrogenation creates (29)_____, which pose a risk to the health of the heart and arteries by raising LDL and lowering HDL cholesterol.

Fats added to foods during (30)_____ or at the table are a major source of fat in the diet. Fats may be (31)_____ on foods, such as the fat trimmed from a steak, or they may be invisible, such as the fat in biscuits or olives. (32)_____ fats are on the rise in U.S. diets.

Chapter Glossary

Matching Exercise:

_____ 1. phospholipids

_____ 2. fats

_____ 3. lipid

_____ 4. oils

_____ 5. cholesterol

_____ 6. emulsifier

_____ 7. sterols

_____ 8. polyunsaturated fatty acid

_____ 9. bile

_____ 10. saturated fatty acid

_____ 11. monounsaturated fatty acid

_____ 12. lecithin

_____ 13. unsaturated fatty acid

_____ 14. monoglycerides

_____ 15. glycerol

_____ 16. triglycerides

_____ 17. cardiovascular disease

_____ 18. saturated fats

a. lipids that are solid at room temperature (70° F or 25° C)

b. an emulsifier made by the liver from cholesterol and stored in the gallbladder

c. one of the three main classes of dietary lipids and the chief form of fat in foods and in the human body

d. a fatty acid containing one point of unsaturation

e. a substance that mixes with both fat and water and permanently disperses the fat in the water, forming an emulsion

f. one of the three main classes of dietary lipids; these lipids are similar to triglycerides, but each has a phosphorus containing acid in place of one of the fatty acids

g. lipids that are liquid at room temperature (70°F or 25°C)

h. a fatty acid carrying the maximum possible number of hydrogen atoms

i. products of the digestion of lipids

j. a phospholipid manufactured by the liver and also found in many foods; a major constituent of cell membranes

k. a family of organic compounds soluble in organic solvents but not in water

l. a fatty acid that lacks some hydrogen atoms and has one or more points of unsaturation

m. a fatty acid with two or more points of unsaturation

n. an organic compound, three carbons long, of interest here because it serves as the backbone for triglycerides

o. one of the three main classes of dietary lipids, with a structure similar to that of cholesterol

p. a member of the group of lipids known as sterols; a soft waxy substance made in the body for a variety of purposes and also found in animal derived foods

q. triglycerides in which most of the fatty acids are saturated

r. disease of the heart and blood vessels

Crossword Puzzle:

	Word Bank:	
chylomicrons	hydrogenation	oxidation
dietary antioxidant	LDL	satiety
essential fatty acids	lipoproteins	smoking point
fatty acids	omega-six fatty acid	*trans* fatty acids
HDL	omega-three fatty acid	

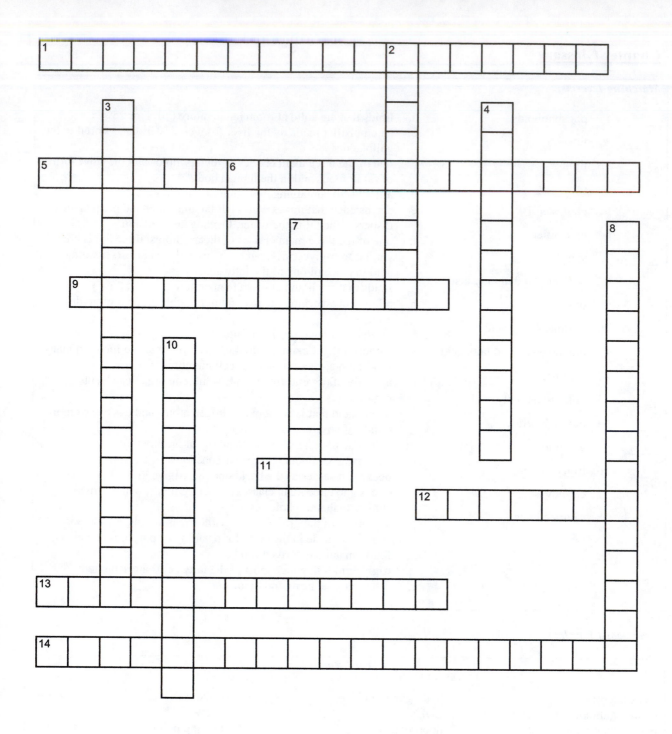

Across:	Down:
1. a substance in food that significantly decreases the damaging effects of reactive compounds, such as reactive forms of oxygen and nitrogen, on tissue functioning	2. interaction of a compound with oxygen
	3. a polyunsaturated fatty acid with its endmost double bond six carbons from the end of the carbon chain
5. a polyunsaturated fatty acid with its endmost double bond three carbons from the end of its carbon chain	4. clusters formed when lipids from a meal are combined with carrier proteins in the intestinal lining
9. clusters of lipids associated with protein, which serve as transport vehicles for lipids in blood and lymph	6. lipoproteins containing a large proportion of protein that return cholesterol from tissues to the liver for dismantling and disposal
11. lipoproteins containing a large proportion of cholesterol that transport lipids from the liver to other tissues such as muscle and fat	7. organic acids composed of carbon chains of various lengths
12. the feeling of fullness or satisfaction that people experience after meals	8. fatty acids with unusual shapes that can arise when polyunsaturated oils are hydrogenated
13. the process of adding hydrogen to unsaturated fatty acids to make fat more solid and resistant to the chemical change of oxidation	10. the temperature at which fat gives off an acrid blue gas
14. fatty acids that the body needs but cannot make in amounts sufficient to meet physiological needs	

Exercises

Answer these chapter study questions:

1. Why is fat, rather than the carbohydrate glucose, the body's major form of stored energy?

2. Why are oils vulnerable to rancidity and how can rancidity be retarded?

3. Why do vegetable oils make up most of the added fat in the U.S. diet?

4. Identify the best way to raise HDL and dietary measures to decrease LDL.

5. What is the best way to ensure the right balance between omega-3 and omega-6 intakes?

6. Why are elevated HDL concentrations associated with a low risk of heart disease?

Complete these short answer questions:

1. The three major classes of lipids include:

    a.
    b.
    c.

2. Triglycerides are made up of three _____ units and _____ glycerol unit.

3. Identify three useful functions of fats in foods.

    a.
    b.
    c.

4. Fatty acids may differ from one another in two ways:

   a.
   b.

5. The two fatty acids essential for human beings include:

   a.
   b.

6. Identify two vegetable oils which are saturated fats:

   a.
   b.

7. Three advantages of hydrogenated oils are:

   a.
   b.
   c.

8. The two major types of lipoproteins of concern to health care professionals include:

   a.
   b.

9. The main dietary factors associated with elevated blood cholesterol are high _____ and *trans* fat intakes.

10. These two vegetable oils are particularly rich in monounsaturates:

    a.
    b.

Solve these problems:

1. Using the exchange lists, calculate the number of grams of fat in the following breakfast menu:

Item	*Grams of Fat*
2 strips of crisp bacon	_____
2 eggs scrambled with 2 teaspoons margarine	_____
½ cup oatmeal	_____
1 slice of toast with 1 teaspoon margarine	_____
1 cup 2% milk	_____
Total:	_____

2. Circle all of the words in the label below which alert you to the fat content of the product.

> **INGREDIENTS: *Chicken stock, chicken, wheat flour, corn starch, cream, vegetable oil, salt, chicken fat, water, margarine, monosodium glutamate, soy protein isolate, yeast extract and natural flavoring.***

3. Modify the following recipe for Beef Stroganoff by making substitutions for the high fat ingredients.

Traditional Recipe

1 pound ground beef
½ cup chopped onion
1 small clove garlic, minced
1 jar sliced mushrooms, drained
2 tablespoons butter
2 tablespoons all-purpose flour
8 ounce carton sour cream
¼ cup chili sauce
½ teaspoon Worcestershire sauce
1 teaspoon salt
¼ teaspoon pepper

Modifications:

Answer these controversy questions:

1. What are some of the problems that may accompany low-fat diets?

2. What are five characteristics of the traditional Mediterranean diet thought to be responsible for its health-promoting characteristics?

3. Explain why nuts may lower heart disease risk.

4. What specific mechanisms support the health benefits of consuming olive oil as protection against heart disease?

5. What would you look for on the label when choosing a margarine most likely to support a healthy heart? Why?

Study Aids

Use Table 5-1 on page 140 in your textbook to study the functions of fat in food and in the human body. Study Figure 5-5 on page 145 of your textbook to recognize the fatty acid composition of common food fats.

Sample Test Items

Comprehension Level Items:

1. Which of the following is **not** one of the three main classes of lipids?

   a. triglycerides
   b. lecithin
   c. phospholipids
   d. sterols

2. Lecithin is an example of a:

   a. phospholipid.
   b. triglyceride.
   c. sterol.
   d. fatty acid.

3. Which of the following is **not** a characteristic of fat in food or in the body?

   a. provides energy reserves
   b. protects the body from temperature extremes
   c. helps cushion vital body organs
   d. decreases the flavor of foods
   e. forms the major material of cell membranes

4. A saturated fatty acid is a fatty acid:

   a. possessing an "empty spot."
   b. having two or more points of unsaturation.
   c. carrying the maximum possible number of hydrogen atoms.
   d. a and b
   e. b and c

5. The essential fatty acids include:

   a. linoleic acid.
   b. linolenic acid.
   c. arachidonic acid.
   d. a and b
   e. b and c

6. Which of the following helps prevent spoilage of oils containing unsaturated fatty acids?

   a. Change them chemically by hydrogenation.
   b. Make them more unsaturated.
   c. Add an antioxidant.
   d. a and b
   e. a and c

7. Which of the following cuts of meat is/are lowest in fat content?

   a. loin
   b. round
   c. sirloin
   d. a and b
   e. b and c

8. A person who eats a high-fat diet incurs a greater-than-average risk of developing:

   a. heart disease.
   b. osteoporosis.
   c. cancer.
   d. a and b
   e. a and c

9. Within the body, fats usually travel from place to place mixed with _____ particles.

   a. carbohydrate
   b. protein
   c. glycerol
   d. glycogen

10. Which of the following statements concerning cholesterol is **not** true?

   a. It is the major part of the plaques that narrow the arteries.
   b. It is widespread in the body and necessary to its function.
   c. It is an important sterol in the structure of cell membranes.
   d. It cannot be made by the body and must be consumed in the diet.

11. The main dietary factor associated with elevated blood cholesterol levels is:

    a. high total fat intake.
    b. monounsaturated fats.
    c. dietary cholesterol.
    d. high saturated and *trans* fat intake.

12. Which of the following food groups are considered to be fat free?

    a. fruits
    b. milk
    c. unprocessed vegetables
    d. a and b
    e. a and c

13. One teaspoon of oil or shortening donates _____ calories.

    a. 25
    b. 30
    c. 40
    d. 45

14. Which of the following is a true statement regarding olestra?

    a. It has not been approved by the FDA.
    b. It passes through the digestive tract unabsorbed.
    c. It has no side effects.
    d. It has been proven safe for use with growing children.

15. A blood lipid profile reveals much useful information regarding a person's risk of:

    a. diabetes.
    b. cancer.
    c. heart disease.
    d. hypoglycemia.

16. In most people, saturated and *trans* fats raise blood cholesterol _____ dietary cholesterol.

    a. more than
    b. less than
    c. the same as
    d. much less than

17. Saturation refers to the number of _____ the fatty acid chain is holding.

    a. oxygens
    b. hydrogens
    c. glycerols
    d. nitrogens

18. *Trans* fatty acids act somewhat like _____ fats in the body.

    a. saturated
    b. unsaturated
    c. monounsaturated
    d. polyunsaturated

19. Linolenic acid, EPA and DHA are examples of:

    a. omega-6 fatty acids.
    b. linoleic acid.
    c. polyunsaturated fatty acids.
    d. omega-3 fatty acids.

20. Of the following fats, which is the hardest?

    a. safflower oil
    b. chicken fat
    c. lard
    d. coconut oil

21. All plant oils are less saturated than animal fats.

    a. true
    b. false

22. Most fat digestion takes place in the:

    a. mouth.
    b. stomach.
    c. small intestine.
    d. colon.

23. Which of the following is critical in the process of carrying cholesterol from body cells to the liver for disposal?

    a. HDL
    b. VLDL
    c. LDL
    d. chylomicrons

24. All of the following are major risk factors for CVD **except**:

    a. family history.
    b. increasing age.
    c. physical inactivity.
    d. female gender.

25. Physical activity promotes heart health because:

    a. it raises blood LDL concentration.
    b. it raises blood HDL concentration.
    c. a smaller volume of blood is pumped with each heartbeat, reducing the heart's workload.
    d. it decreases circulation.

26. A diet that provides _____% of its calories from fat is currently recommended by health authorities.

    a. 10-25
    b. 20-35
    c. 25-40
    d. 30-45

Application Level Items:

27. Someone asks you whether he should reduce dietary cholesterol in an effort to prevent cardiovascular disease. An appropriate response would be:

    a. cholesterol, in general, doesn't matter.
    b. it doesn't matter as much as saturated and *trans* fat intake.
    c. dietary cholesterol makes a major contribution.
    d. very few foods contain cholesterol anyway.

28. You are trying to get a group of people to meet the current dietary recommendations for fats. You tell them that one of the most effective steps they can take at home is to:

    a. replace butter with margarine.
    b. eat fewer foods.
    c. eliminate solid, saturated fats used as seasonings.
    d. consume fewer snacks throughout the day.

29. Research suggests that North Americans should consume more omega-3 fatty acids. In order to do this, you would consume more:

    a. beef.
    b. safflower oil.
    c. corn oil.
    d. salmon.

30. You are trying to convince a friend not to buy fish oil supplements. You would use all of the following as arguments **except**:

    a. concentrated supplements make it easy to overdose on toxic vitamins.
    b. the supplements are very expensive.
    c. overdoses may cause heart disease and cancer.
    d. supplements may have toxic concentrations of pesticides.

31. A man's father and grandfather both died of heart disease. After ordering a blood lipid profile, the man's physician told him his LDL was too high and his HDL too low. He sometimes walks for exercise, but he smokes cigarettes daily. Would this person be considered at risk for heart disease?

    a. yes
    b. no

32. Which of the following would provide the least amount of *trans* fatty acids?

    a. peanut butter
    b. blue cheese salad dressing
    c. liquid margarine
    d. potato chips

Answers

Summing Up

(1) three; (2) triglycerides; (3) sterols; (4) harmful; (5) energy; (6) organs; (7) temperature; (8) fatty acids; (9) K; (10) triglycerides; (11) chain; (12) saturation; (13) liquid; (14) fish; (15) palm; (16) lipoproteins; (17) protein; (18) LDL; (19) HDL; (20) low; (21) cholesterol; (22) saturated; (23) linolenic; (24) essential; (25) oxygen; (26) rancid; (27) air; (28) hydrogenation; (29) *trans* fatty acids; (30) preparation; (31) visible; (32) invisible

Chapter Glossary

Matching Exercise: (1) f; (2) a; (3) k; (4) g; (5) p; (6) e; (7) o; (8) m; (9) b; (10) h; (11) d; (12) j; (13) l; (14) i; (15) n; (16) c ; (17) r; (18) q

Crossword Puzzle: (1) dietary antioxidant; (2) oxidation; (3) omega-6 fatty acid; (4) chylomicrons; (5) omega-3 fatty acid; (6) HDL; (7) fatty acids; (8) *trans* fatty acids; (9) lipoproteins; (10) smoking point; (11) LDL; (12) satiety; (13) hydrogenation; (14) essential fatty acids

Exercises

Chapter Study Questions:

1. Glycogen, the stored form of carbohydrate, holds a great deal of water and is quite bulky. Therefore, the body cannot store enough glycogen to provide energy for very long. In contrast, fats pack tightly together without water and can store many more calories of energy in a small space.

2. Oils contain points of unsaturation in the fatty acids and these are vulnerable to attack by oxygen. When the unsaturated points are oxidized, the oils become rancid. To prevent rancidity, oils should be stored in tightly covered containers that exclude air and they should be placed in the refrigerator if stored for long periods of time.

3. Because fast food chains use them for frying, manufacturers add them to processed foods, and consumers tend to choose margarine over butter.

4. To raise HDL, people should exercise more; to decrease LDL, people should replace saturated fat with monounsaturated or polyunsaturated fats.

5. Eat meals of fish 2-3 times a week, total; about 10 ounces of fish per week.

6. Because of the task they perform in scavenging excess cholesterol and phospholipids from the body's tissues for disposal.

Short Answer Questions:

1. (a) triglycerides; (b) phospholipids; (c) sterols

2. fatty acid; one

3. (a) provide a concentrated energy source in foods; (b) enhance food's aroma and flavor; (c) contribute to satiety. Refer to Table 5-1 for additional functions of fats in food.

4. (a) in chain length; (b) in degree of saturation

5. (a) linoleic; (b) linolenic

6. (a) coconut oil; (b) palm oil

7. (a) resist rancidity; (b) firm texture; (c) have a high smoking point

8. (a) LDL; (b) HDL

9. saturated

10. (a) olive; (b) canola

Problem-Solving:

1.

Item	Grams of Fat
2 strips of crisp bacon	10
2 eggs scrambled with 2 teaspoons margarine	20
½ cup oatmeal	0-1
1 slice of toast with 1 teaspoon margarine	5-6
1 cup 2% milk	5
Total:	**40-42**

2. Chicken stock; chicken; cream; vegetable oil; chicken fat; margarine

3. Use round steak instead of ground beef; use reduced calorie margarine in the place of butter; use yogurt instead of sour cream.

Controversy Questions:

1. They are difficult to maintain over time; they are not necessarily low-calorie diets; may exclude nutritious foods that provide essential fatty acids, phytochemicals and other nutrients.

2. The diet is low in saturated fat, very low in *trans* fat, rich in unsaturated fat, rich in complex carbohydrates and fiber, and rich in nutrients and phytochemicals that support good health.

3. Nuts are low in saturated fats, high in fiber, vegetable protein and the antioxidant vitamin E, and they are high in phytochemicals that act as antioxidants.

4. Olive oil may conserve the beneficial blood lipid, HDL, while lowering levels of the riskier types of blood lipids; it contains phytochemicals that act as antioxidants; it also lowers LDL cholesterol vulnerability to oxidation, lowers blood clotting factors, and lowers blood pressure.

5. Look for oil as the first ingredient because that means it is probably low in saturated fat and *trans* fatty acids.

Sample Test Items

1. b (p. 140)	9. b (p. 147)	17. b (p. 143)	25. b (p. 154)
2. a (p. 140)	10. d (p. 145-146)	18. a (p. 159)	26. b (p. 150)
3. d (p. 140-142)	11. d (p. 151)	19. d (p. 155)	27. b (p. 151)
4. c (p. 143)	12. e (p. 161, 164-168)	20. c (p. 143, 145)	28. c (p. 168)
5. d (p. 154)	13. d (p. 161)	21. b (p. 143-144)	29. d (p. 155-156)
6. e (p. 158-159)	14. b (p. 162-163)	22. c (p. 146-147)	30. c (p. 157)
7. d (p. 164)	15. c (p. 150-151)	23. a (p. 149)	31. a (p. 151)
8. e (p. 149)	16. a (p. 151)	24. d (p. 151)	32. c (p. 158, 160)

Chapter 6 – The Proteins and Amino Acids

Chapter Objectives

After completing this chapter, you should be able to:

1. Describe the structure of proteins and explain why adequate amounts of all the essential amino acids are required for protein synthesis.

2. Summarize the processes of protein digestion and absorption in the body.

3. Explain the roles of protein in the body.

4. Discuss the importance of protein quality and mutual supplementation.

5. Calculate your individual protein RDA and describe how nitrogen balance studies are used in determining the RDA for protein.

6. Describe the consequences of both protein deficiency and protein excess.

7. Compare the positive health aspects of a vegetarian diet with those of a diet that includes meat and describe ways each diet can include adequate nutrients (Controversy 6).

Key Concepts

✓ Proteins are unique among the energy nutrients in that they possess nitrogen-containing amine groups and are composed of 20 different amino acid units. Of the 20 amino acids, some are essential, and some are essential only in special circumstances.

✓ Amino acids link into long strands that coil and fold to make a wide variety of different proteins.

✓ Each type of protein has a distinctive sequence of amino acids and so has great specificity. Often, cells specialize in synthesizing particular types of proteins in addition to the proteins necessary to all cells. Nutrients act as environmental signals, affecting genetic activities.

✓ Proteins can be denatured by heat, acids, bases, alcohol, or the salts of heavy metals. Denaturation begins the process of digesting food protein and can also destroy body proteins.

✓ Digestion of protein involves denaturation by stomach acid, then enzymatic digestion in the stomach and small intestine to amino acids, dipeptides, and tripeptides.

✓ The cells of the small intestine complete digestion, absorb amino acids and some larger peptides, and release them into the bloodstream for use by the body's cells.

✓ The body needs dietary amino acids to grow new cells and to replace worn-out ones.

✓ The body makes enzymes, hormones, and chemical messengers of the nervous system from its amino acids.

✓ Antibodies are proteins that defend against foreign proteins and other foreign substances within the body.

✓ Proteins help to regulate the body's electrolytes and fluids.

✓ Proteins buffer the blood against excess acidity or alkalinity.

✓ Proteins that clot the blood prevent uncontrolled bleeding from injuries.

✓ When insufficient carbohydrate and fat are consumed to meet the body's energy need, food protein and body protein are sacrificed to supply energy. The nitrogen part is removed from each amino acid, and the resulting fragment is oxidized for energy. No storage form of amino acids exists in the body.

✓ Amino acids can be metabolized to protein, nitrogen plus energy, glucose, or fat. They will be metabolized to protein only if sufficient energy is present from other sources. The diet should supply all essential amino acids and a full measure of protein according to guidelines.

✓ The body's use of a protein depends in part on the user's health, the protein quality, and the other nutrients and energy taken with it. Digestibility of protein varies from food to food and cooking can improve or impair it.

✓ A protein's amino acid assortment greatly influences its usefulness to the body. Proteins lacking needed amino acids can be used only if those amino acids are present from other sources.

✓ The quality of a protein is measured by its amino acids, by its digestibility, and by how well it meets human needs. The protein value listed on a food label reflects the PDCAAS of the protein in the food.

✓ The amount of protein needed daily depends on size and stage of growth. The DRI recommended intake for adults is 0.8 grams of protein per kilogram of body weight. Protein recommendations are based on nitrogen balance studies, which compare nitrogen excreted from the body with nitrogen ingested in food.

✓ Protein-deficiency symptoms are always observed when either protein or energy is deficient. Extreme food energy deficiency is marasmus; extreme protein deficiency is kwashiorkor. The two diseases overlap most of the time, and together are called PEM.

✓ Health risks may follow the overconsumption of protein-rich foods.

Summing Up

Protein is a compound composed of carbon, hydrogen, oxygen, and (1)_____ atoms. Protein is made up of building blocks called (2)_____. About (3)_____ amino acids, each with its different side chain, make up most of the proteins of living tissue. The side chains make the amino acids differ in size, (4)_____, and electrical charge.

There are some amino acids that the healthy adult body makes too slowly or cannot make at all and these are called the (5)_____ amino acids. In the first step of making a protein, each amino acid is hooked to the next. A chemical bond, called a (6)_____ bond, is formed between the amine group end of one and the (7)_____ group end of the next. The (8)_____ of proteins enable them to perform different tasks in the body. Among the most fascinating of the proteins are the (9)_____, which act on other substances to change them chemically.

Proteins can be denatured by heat, alcohol, (10)_____, bases, or the salts of heavy metals. During digestion of proteins, an early step is denaturation by the (11)_____. In the stomach, the (12)_____ helps to uncoil the protein's strands so that the stomach enzymes can attack the peptide bonds. The stomach lining, which is made partly of (13)_____, is protected by a coat of (14)_____. Continual digestion, and finally absorption, of protein takes place in the (15)_____.

Protein serves many roles and the body needs dietary amino acids to grow new (16)_____ and to maintain fluid and electrolyte balance and (17)_____ balance. Under conditions of inadequate energy or (18)_____, protein will be sacrificed to provide needed (19)_____. When amino acids are degraded for energy, their (20)_____ groups are stripped off and used elsewhere or are incorporated by the liver into (21)_____ to be excreted in the urine.

Generally, amino acids from (22)_____ proteins are best absorbed. If the diet fails to provide enough of an (23)_____ amino acid, the cells begin to adjust their activities almost immediately. The (24)_____ is the most important method of evaluating the quality of food protein and is used by those establishing protein values on food labels. On the PDCAAS scale of 100 to 0, with 100 representing protein sources that are most readily (25)_____ and most perfectly balanced, (26)_____, ground beef, chicken products, fat-free milk and (27)_____ all score 100. (28)_____ protein isn't far behind at 94.

The DRI for protein depends on (29)_____ and it is higher for infants and (30)_____ and pregnant and lactating women to cover needs for building new tissue. For healthy adults, the DRI recommended intake for protein has been set at (31)_____ grams for each kilogram of body weight.

Two groups of foods in the USDA Food Guide contribute high-quality protein, including the (32)_____ group and the meat group. In addition, the vegetable and (33)_____ groups contribute smaller amounts of proteins. Protein-rich foods carry with them a characteristic array of vitamins and (34)_____, including iron, but they lack others, such as vitamin C and (35)_____.

Chapter Glossary

Matching Exercise:

_____ 1. immunity

_____ 2. peptide bond

_____ 3. polypeptides

_____ 4. conditionally essential amino acid

_____ 5. bases

_____ 6. edema

_____ 7. dipeptides

_____ 8. hormones

_____ 9. enzymes

_____ 10. amino acid pools

_____ 11. acid-base balance

_____ 12. fluid and electrolyte balance

_____ 13. protein turnover

_____ 14. acids

_____ 15. hemoglobin

_____ 16. protein-energy malnutrition

_____ 17. denaturation

_____ 18. antibodies

a. chemical messengers secreted by a number of body organs in response to conditions that require regulation

b. the continuous breakdown and synthesis of body proteins involving recycling of amino acids

c. protein fragments of many (more than ten) amino acids bonded together

d. large proteins of the blood, produced by the immune system in response to invasion of the body by foreign substances (antigens)

e. protection from or resistance to a disease or infection by development of antibodies and by the actions of cells and tissues in response to a threat

f. amino acids dissolved in the body's fluids that provide cells with ready raw materials from which to build new proteins or other molecules

g. the world's most widespread malnutrition problem, including both marasmus and kwashiorkor and states in which they overlap

h. swelling of body tissue caused by leakage of fluid from the blood vessels

i. distribution of fluid and dissolved particles among body compartments

j. an amino acid that is normally nonessential, but must be supplied by the diet in special circumstances when the need for it exceeds the body's ability to produce it

k. the globular protein of red blood cells whose iron atoms carry oxygen around the body

l. a bond that connects one amino acid with another, forming a link in a protein chain

m. compounds that release hydrogens in a watery solution

n. protein catalysts

o. the change in a protein's shape brought about by heat, acids, bases, alcohol, salts of heavy metals, or other agents

p. equilibrium between acid and base concentrations in the body fluids

q. protein fragments that are two amino acids long

r. compounds that accept hydrogens from solutions

Crossword Puzzle:

Word Bank:			
acidosis	complementary proteins	limiting amino acid	serotonin
alkalosis	dysentery	marasmus	textured vegetable protein
buffers	kwashiorkor	mutual supplementation	tofu
collagen	legumes	PDCAAS	urea

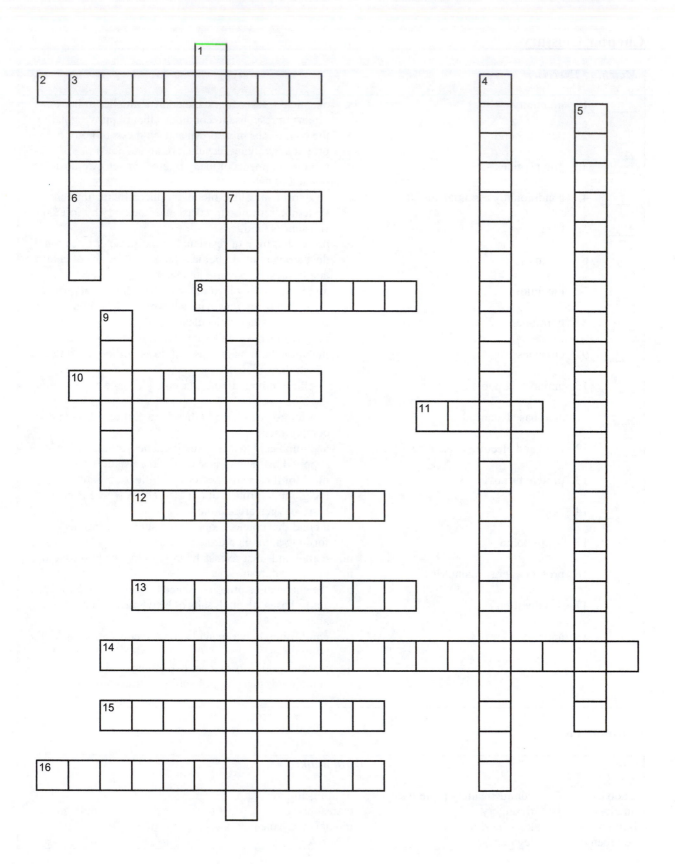

Crossword Puzzle Across:	Crossword Puzzle Down:
2. blood alkalinity above normal	1. a curd made from soybean that is rich in protein, often rich in calcium and variable in fat content
6. the calorie-deficiency disease; starvation	3. plants of the bean, pea, and lentil family that have roots with nodules containing special bacteria
8. compounds that help keep a solution's acidity or alkalinity constant	4. processed soybean protein used in products formulated to look and taste like meat, fish, or poultry
10. blood acidity above normal, indicating excess acid	5. two or more proteins whose amino acid assortments complement each other in such a way that the essential amino acids missing from one are supplied by the other
11. the principal nitrogen-excretion product of metabolism, generated mostly by removal of amine groups from unneeded amino acids or from amino acids being sacrificed to a need for energy	7. the strategy of combining two incomplete protein sources so that the amino acids in one food make up for those lacking in the other food
12. a type of body protein from which connective tissues such as scars, tendons, ligaments, and the foundations of bones and teeth are made	9. a measuring tool used to determine protein quality; reflects a protein's digestibility as well as the proportions of amino acids that it provides
13. an infection of the digestive tract that causes diarrhea	
14. an essential amino acid present in dietary protein in an insufficient amount, so that it limits the body's ability to build protein	
15. a compound related in structure to (and made from) the amino acid tryptophan	
16. a disease related to protein malnutrition, with a set of recognizable symptoms, such as edema	

Exercises

Answer these chapter study questions:

1. Why is milk given as a first-aid remedy when someone swallows a heavy-metal poison?

2. Describe the conditions under which amino acids would be wasted (i.e., not used to build protein or other nitrogen-containing compounds).

3. Describe the main difference between the dietary inadequacies responsible for marasmus versus kwashiorkor.

4. What are legumes and why are they especially protein-rich food sources?

5. Why is it nutritionally advantageous to cook an egg?

6. Why does the DRI Committee recommend that the diet contain no more than 35% of calories from protein?

Complete these short answer questions:

1. Proteins contain the following types of atoms:

   a.
   b.
   c.
   d.

2. The nine essential amino acids for adults include:

   a.   f.
   b.   g.
   c.   h.
   d.   i.
   e.

3. Amino acids in the cell can be:

   a.
   b.
   c.
   d.
   e.

4. Protein can undergo denaturation by:

   a.
   b.
   c.
   d.
   e.

5. Examples of proteins that contain ample amounts of all the essential amino acids include:

   a.   e.
   b.   f.
   c.   g.
   d.

6. Examples of individuals in positive nitrogen balance include:

   a.
   b.

7. List the steps necessary to calculate a person's DRI for protein.

   a.
   b.
   c.

8. The food groups in the USDA Food Guide which contribute protein to the diet include:

   a.
   b.
   c.
   d.

9. Protein-rich foods are notoriously lacking in the nutrients _____ and _____.

10. Animal protein food sources are high in vitamin _____ and the mineral _____.

Solve these problems:

1. What is the DRI for protein for a 37-year-old female who is 5'4" tall and weighs 110 pounds?

2. Calculate the number of grams of protein provided in the following menu, using the U.S. Food Exchange System found in Appendix D of your textbook.

Menu Item	Grams of Protein
Breakfast	
1 cup skim milk	_____
1 boiled egg	_____
1 slice whole wheat toast	_____
Lunch	
3 oz. hamburger patty	_____
1 whole wheat bun	_____
tomato and lettuce (1 cup total)	_____
2 tsp. mayonnaise	_____
Dinner	
3 oz. chicken breast	_____
½ cup green beans	_____
1 roll	_____
1 cup skim milk	_____
Total:	_____

Answer these controversy questions:

1. What are some of the different reasons why individuals are vegetarians?

2. What are five conditions that a vegetarian diet may offer protection against?

3. Other than food choices, how do vegetarians differ from non-vegetarians?

4. Identify the dietary constituents that, when consumed in moderate to high amounts, are correlated with colon cancer.

5. Which nutrients require special attention for strict vegetarians?

Study Aids

1. Identify the number of servings recommended for each food group in the Vegetarian Food Pyramid (p. 211).

Food Group	_Number of Servings_
a. Bread, pasta, rice, fortified cereals	_____
b. Green leafy vegetables	_____
c. Vegetables	_____
d. Fruits	_____
e. Dried fruits	_____
f. Beans and protein foods	_____
g. Nuts and seeds	_____

2. Rank the following food items according to their PDCAAS values by placing a one next to the item with the highest score, a two next to the item with the next highest score, etc.

_____ soybean protein
_____ wheat protein
_____ egg white

3. Use Table 6-1 on page 191 of your textbook to study the functions of protein in the body. Study Figure 6-12 on page 192 of your textbook to compare and contrast the three different energy sources of carbohydrate, protein, and fat.

Sample Test Items

Comprehension Level Items:

1. A strand of amino acids that makes up a protein may contain _____ different kinds of amino acids.

   a. 5
   b. 10
   c. 15
   d. 20

2. Which of the following makes amino acids differ in size, shape, and electrical charge?

   a. the side chains
   b. the amine groups
   c. the acid groups
   d. a and b
   e. b and c

3. Which of the following is **not** considered to be an essential amino acid?

   a. threonine
   b. alanine
   c. lysine
   d. phenylalanine

4. The body does **not** make a specialized storage form of protein as it does for carbohydrate and fat.

   a. true
   b. false

5. Appropriate roles of protein in the body include all of the following **except**:

   a. to provide the body's needed energy.
   b. for growth and development.
   c. to maintain fluid and electrolyte balance.
   d. for acid-base balance.

6. When there is a surplus of amino acids and energy-yielding nutrients, the body:

   a. removes and excretes their amine groups.
   b. uses them to build extra muscle tissue.
   c. converts them to fat for storage.
   d. a and b
   e. a and c

7. Athletes need slightly more protein than the DRI, but the increased need is well covered by a regular diet.

   a. true
   b. false

8. Which of the following methods of evaluating protein quality is used as a standard for the majority of food labeling?

    a. chemical score
    b. net protein utilization
    c. protein efficiency ratio
    d. protein digestibility-corrected amino acid score

9. Of the following foods, which contains amino acids that are best digested and absorbed?

    a. whole wheat bread
    b. chicken
    c. legumes
    d. broccoli

10. The calorie deficiency disease is known as:

    a. protein-energy malnutrition.
    b. kwashiorkor.
    c. protein-calorie malnutrition.
    d. marasmus.

11. Characteristics of children with marasmus include all of the following **except**:

    a. their digestive enzymes are in short supply.
    b. they retain some of their stores of body fat.
    c. all of their muscles are wasted.
    d. their metabolism is slowed.

12. Under normal circumstances, healthy adults are in:

    a. nitrogen equilibrium.
    b. positive nitrogen balance.
    c. negative nitrogen balance.
    d. minus zero balance.

13. Which of the following is in positive nitrogen balance?

    a. a growing child
    b. a person who rests in bed for a long time
    c. a pregnant woman
    d. a and b
    e. a and c

14. High-protein diets are associated with all of the following **except**:

    a. high-fat foods.
    b. increased risk of obesity.
    c. improved kidney function.
    d. increased risk of heart disease.

15. Health recommendations typically advise a protein intake between 10 and 35% of energy intake.

    a. true
    b. false

16. Which of the following is **not** characteristic of legumes?

    a. They have nodules on their roots containing bacteria that can fix nitrogen.
    b. They capture nitrogen from the air and soil and use it to make amino acids.
    c. They are used by farmers in rotation with other crops to fertilize fields.
    d. Their protein quality is inferior to that of other plant foods.

17. Legumes are excellent sources of all of the following nutrients **except**:

    a. iron.
    b. calcium.
    c. vitamin C.
    d. fiber.

18. Tofu is:

    a. naturally rich in calcium.
    b. a curd made from soybeans.
    c. a poor source of protein.
    d. high in fiber.

19. These food groups contribute an abundance of high-quality protein:

    a. meat, poultry, fish, legumes, eggs and nuts.
    b. grains.
    c. milk, yogurt and cheese.
    d. a and b
    e. a and c

20. Under special circumstances, a nonessential amino acid can become essential.

    a. true
    b. false

21. The first step in the destruction of a protein is called:

    a. protein turnover.
    b. net protein utilization.
    c. denaturation.
    d. protein efficiency ratio.

22. A careful review of literature concluded that enough research exists to support recommending long-term consumption of amino acid supplements by healthy people.

    a. true
    b. false

23. The millions of cells lining the intestinal tract live for _____ days.

    a.   1
    b.   2
    c.   3
    d.   4

24. The amino acid tryptophan serves as starting material for:

    a.   the neurotransmitter serotonin.
    b.   the hormone thyroxine.
    c.   the vitamin niacin.
    d.   a and b
    e.   a and c

Application Level Items:

25. Your best friend is a body builder who puts raw eggs in a milkshake to increase his protein intake. What argument would you use against this practice?

    a.   Raw egg proteins bind the B vitamin biotin.
    b.   Raw egg proteins speed up protein digestion.
    c.   Raw egg proteins bind the mineral iron.
    d.   a and b
    e.   a and c

26. A person who centers his diet around protein is likely to receive an inadequate intake of:

    a.   vitamin B_{12}.
    b.   fat.
    c.   folate.
    d.   iron.

27. Valuable, expensive protein-rich foods can contribute to obesity when:

    a.   there is a surplus of amino acids.
    b.   they are used as is and become part of a growing protein.
    c.   there is a surplus of energy-yielding nutrients.
    d.   a and c
    e.   b and c

28. Which of the following has the highest DRI for protein per unit of body weight?

    a.   a 6-month-old infant
    b.   a healthy 25-year-old female
    c.   a 35-year-old male
    d.   a 30-year-old female who exercises frequently

29. What advice would you give to a friend who wants to build bigger muscles?

    a. Consume extra protein in the form of foods.
    b. Engage in rigorous physical training.
    c. Take an amino acid supplement.
    d. a and b
    e. b and c

30. Approximately 7 grams of protein are provided by each of the following foods, which would be considered as cooked and/or ready to eat: **½ cup legumes; 1 cup broccoli; 1 ounce cheese; 2 tablespoons peanut butter.** Which of these would be the most nutrient-dense food choice?

    a. 1 cup broccoli
    b. 2 tablespoons peanut butter
    c. 1 ounce cheese
    d. ½ cup legumes

31. What is the DRI for protein for a 50-year-old male weighing 198 pounds?

    a. 56
    b. 67
    c. 72
    d. 80

Answers

Summing Up

(1) nitrogen; (2) amino acids; (3) twenty; (4) shape; (5) essential; (6) peptide; (7) acid; (8) shapes; (9) enzymes; (10) acids; (11) stomach acid; (12) acid; (13) protein; (14) mucus; (15) small intestine; (16) cells; (17) acid-base; (18) carbohydrate; (19) energy; (20) amine; (21) urea; (22) animal; (23) essential; (24) PDCAAS; (25) digested; (26) egg white; (27) tuna fish; (28) soybean; (29) body weight; (30) children; (31) 0.8; (32) milk; (33) grain; (34) minerals; (35) folate

Chapter Glossary

Matching Exercise: (1) e; (2) l; (3) c; (4) j; (5) r; (6) h; (7) q; (8) a; (9) n; (10) f; (11) p; (12) i; (13) b; (14) m; (15) k; (16) g; (17) o; (18) d
Crossword Puzzle: (1) tofu; (2) alkalosis; (3) legumes; (4) textured vegetable protein; (5) complementary proteins; (6) marasmus; (7) mutual supplementation; (8) buffers; (9) PDCAAS; (10) acidosis; (11) urea; (12) collagen; (13) dysentery; (14) limiting amino acid; (15) serotonin; (16) kwashiorkor

Exercises

Chapter Study Questions:

1. Many poisons are salts of heavy metals which denature proteins whenever they touch them. Milk is given so that the poison will act on the protein of the milk rather than on the protein tissues of the mouth, esophagus, and stomach.

2. Whenever there is not enough energy from carbohydrate or fat; when the diet's protein is low quality, with too few essential amino acids; when there is too much protein so that not all is needed; when there is too much of any amino acid, such as from a supplement.

3. Marasmus reflects a chronic inadequate food intake and inadequate energy, vitamins, and minerals as well as too little protein; kwashiorkor results from severe child malnutrition with too little protein to support body functions.

4. Legumes are plants of the pea, bean and lentil family which have roots with nodules that contain special bacteria. These bacteria can trap nitrogen from the air in the soil and make it into compounds that become part of the seed. The seeds are rich in high-quality protein compared with those of most other plants.

5. Cooking an egg denatures the protein and makes it firm; in addition, cooking denatures two raw-egg proteins that bind the B vitamin biotin and the mineral iron and slows protein digestion; therefore, cooking eggs liberates biotin and iron and aids digestion.

6. Because overconsumption of protein offers no benefits and may be related to health problems related to the heart, kidneys, and bones.

Short Answer Questions:

1. (a) oxygen; (b) hydrogen; (c) carbon; (d) nitrogen

2. (a) valine; (b) leucine; (c) isoleucine; (d) threonine; (e) lysine; (f) methionine; (g) phenylalanine; (h) tryptophan; (i) histidine

3. (a) used to build proteins; (b) converted to other small nitrogen-containing compounds such as niacin; (c) converted to some other amino acids; (d) converted to glucose or fat; (e) burned as fuel

4. (a) heat; (b) alcohol; (c) acids; (d) bases; (e) salts of heavy metals

5. (a) meat; (b) fish; (c) poultry; (d) cheese; (e) eggs; (f) milk; (g) many soybean products

6. (a) growing child; (b) pregnant woman

7. (a) find body weight in pounds; (b) convert pounds to kilograms; (c) multiply by 0.8 g/kg to find total grams of protein per day

8. (a) meat; (b) milk; (c) vegetable; (d) grain

9. vitamin C; folate

10. vitamin B_{12}; iron

Problem-Solving:

1. 40 grams protein

2.

Menu Item	***Grams of Protein***
Breakfast	
1 cup skim milk	8
1 boiled egg	7
1 slice whole wheat toast	3
Lunch	
3 oz. hamburger patty	21
1 whole wheat bun	6
tomato and lettuce (1 vegetable serving)	2
2 tsp. mayonnaise	

<u>*Dinner*</u>
3 oz. chicken breast	21
½ cup green beans	2
1 roll	3
1 cup skim milk	8

Total: **81**

Controversy Questions:

1. Some people do not believe in killing animals for their meat and believe that livestock is treated inhumanely. Others fear contracting diseases, such as mad cow disease, or practice vegetarianism for health reasons.

2. Obesity, high blood pressure, heart disease, and some forms of cancer.

3. Vegetarians typically use no tobacco, use alcohol in moderation and may be more physically active.

4. Alcohol, total food energy, fatty red meats and processed meats, and possibly high intakes of refined grain products.

5. Calcium, iron, zinc, vitamin D and vitamin B_{12} all require special attention for strict vegetarians.

Study Aids

1. (a) 6-10; (b) 2-3; (c) 2-4; (d) 1-2; (e) 1-2; (f) 2-3; (g) 1-2

2.
2	soybean protein
3	wheat protein
1	egg white

Sample Test Items

1. d (p. 180)	12. a (p. 197)	22. b (p. 193)
2. a (p. 180)	13. e (p. 197)	23. c (p. 188)
3. b (p. 181)	14. c (p. 201)	24. e (p. 188)
4. a (p. 191)	15. a (p. 202)	25. e (p. 184)
5. a (p. 188-191)	16. d (p. 202, 205)	26. c (p. 202)
6. e (p. 192)	17. c (p. 202)	27. d (p. 191)
7. a (p. 196)	18. b (p. 204)	28. a (p. 196)
8. d (p. 196)	19. e (p. 202)	29. b (p. 184)
9. b (p. 194)	20. a (p. 180)	30. a (p. 203)
10. d (p. 199)	21. c (p. 184)	31. c (p. 196)
11. b (p. 199)		

Chapter 7 - The Vitamins

Chapter Objectives

After completing this chapter, you should be able to:

1. Define the term vitamin and explain how vitamins are classified.

2. Describe the characteristics of fat-soluble vitamins and water-soluble vitamins and explain how they differ.

3. List the chief functions and food sources of each vitamin and describe any major deficiency and toxicity symptoms associated with each.

4. Identify groups of people who may benefit from a multinutrient supplement and discuss guidelines for choosing an appropriate supplement.

5. Describe the best method of planning a diet that is rich in vitamins.

6. Explain how supplements are regulated based on the Dietary Supplement Health and Education Act, and discuss evidence regarding the pros and cons of vitamin supplements (Controversy 7).

Key Concepts

✓ Vitamins are essential, noncaloric nutrients that are needed in tiny amounts in the diet and help to drive cell processes in the body. The fat-soluble vitamins are vitamins A, D, E, and K; the water-soluble vitamins are vitamin C and the B vitamins.

✓ Vitamin A is essential to vision, integrity of epithelial tissue, bone growth, reproduction, and more. Vitamin A deficiency causes blindness, sickness, and death and is a major problem worldwide. Overdoses are possible and cause many serious symptoms. Foods are preferable to supplements for supplying vitamin A.

✓ The vitamin A precursor in plants, beta-carotene, is an effective antioxidant in the body. Brightly colored plant foods are richest in beta-carotene, and diets containing these foods are associated with eye health.

✓ Vitamin D raises mineral levels in the blood, notably calcium and phosphorus, permitting bone formation and maintenance. A deficiency can cause rickets in childhood or osteomalacia in later life. Vitamin D is the most toxic of all the vitamins, and excesses are dangerous or deadly. People exposed to the sun make vitamin D from a cholesterol-like compound in their skin; fortified milk is an important food source.

✓ Vitamin E acts as an antioxidant in cell membranes and is especially important for the integrity of cells that are constantly exposed to high oxygen concentrations, namely, the lungs and blood cells, both red and white. Vitamin E deficiency is rare in human beings, but it does occur in newborn premature infants. The vitamin is widely distributed in plant foods; it is destroyed by high heat; toxicity is rare.

✓ Vitamin K is necessary for blood to clot; deficiency causes uncontrolled bleeding. The bacterial inhabitants of the digestive tract produce vitamin K. Toxicity causes jaundice.

✓ Vitamin C, an antioxidant, helps to maintain collagen, the protein of connective tissue; protects against infection; and helps in iron absorption. The theory that vitamin C prevents or cures colds or cancer is not well supported by research. Taking high vitamin C doses may be unwise. Ample vitamin C can be obtained from foods.

✓ As part of coenzymes, the B vitamins help enzymes do their jobs. The B vitamins facilitate the work of every cell. Some help generate energy; others help make protein and new cells. B vitamins work everywhere in the body tissue to metabolize carbohydrate, fat, and protein.

✓ Historically, famous B vitamin-deficiency diseases are beriberi (thiamin), pellagra (niacin), and pernicious anemia (vitamin B_{12}). Pellagra can be prevented by adequate protein because the amino acid tryptophan can be converted to niacin in the body. A high intake of folate can mask the blood symptom of vitamin B_{12} deficiency but will not prevent the associated nerve damage. Vitamin B_6 is important in amino acid metabolism and can be toxic in excess. Biotin and pantothenic acid are important to the body and are abundant in food.

✓ Choline is needed in the diet, but it is not a vitamin and deficiencies are unheard of outside the laboratory. Many other substances that people claim are B vitamins are not. Among these substances are carnitine, inositol, and lipoic acid.

Summing Up

A vitamin is defined as an essential, noncaloric, organic nutrient needed in (1)_____ amounts in

the diet. The role of many vitamins is to help make possible the processes by which other nutrients are digested,

(2)_____, and metabolized, or built into body (3)_____. Some of the vitamins

occur in foods in a form known as (4)_____, or provitamins. The vitamins fall naturally into two

classes including the (5)_____-soluble vitamins and the (6)_____-soluble vitamins.

Vitamin A plays parts in such diverse functions as (7)_____, maintenance of skin and body

linings, bone growth and reproduction. One of the active forms of vitamin A is (8)_____, which is

stored in the liver. In plants, vitamin A exists only in its precursor forms; the most abundant of these is

(9)_____. Rich food sources of beta carotene include carrots, sweet potatoes, pumpkins, mango,

(10)_____, and apricots, as well as dark green vegetables.

Vitamin D is one member of a team of nutrients and hormones that maintains blood calcium and phosphorus

levels and thereby (11)_____ integrity. The vitamin D deficiency disease in

(12)_____ is called rickets, while the comparable disease in adults is referred to as

(13)_____. Vitamin D is the most potentially (14)_____ of all vitamins and as

little as five times the recommended amount has been associated with signs of Vitamin D toxicity. People can make

vitamin D from a (15)_____ compound whenever the sun shines on their skin. The recommended

intake for vitamin D is (16)_____ micrograms per day for adults 19 to 50 years.

Vitamin E is an (17)_____ and thus serves as one of the body's main defenders against

oxidative damage. In the body, it exerts an especially important antioxidant effect in the

(18)_____. Conditions that cause malabsorption of (19)_____ can cause vitamin E deficiencies, although vitamin E deficiency is rare in most situations. The DRI intake recommendation for vitamin E is (20)_____ milligrams a day for adults. About 20% of the vitamin E people consume comes from (21)_____ oils and products made from them.

Vitamin K is the fat-soluble vitamin necessary for the synthesis of proteins involved in (22)_____. It can be obtained from the nonfood source of (23)_____ bacteria. However, vitamin K deficiencies may occur in newborn infants and people who have taken (24)_____ that have killed their intestinal bacteria.

The B vitamins and vitamin C are known together as the (25)_____ vitamins. The B vitamins act as part of (26)_____. Niacin, thiamin and riboflavin participate in the release of (27)_____ from carbohydrate, fat, and protein. For (28)_____, which is tied closely to amino acid metabolism, the amount needed is proportional to (29)_____ intake. The thiamin deficiency disease is called (30)_____, while the (31)_____ deficiency disease is known as pellagra. In the body, the amino acid (32)_____ is converted to niacin. Strict vegetarians are most apt to be deficient in (33)_____, which is present only in foods of animal origin. The vitamin C deficiency disease is scurvy, which can be prevented by an intake of (34)_____ milligrams of vitamin C per day. However, the adult DRI intake recommendation for vitamin C is (35)_____ milligrams for men and 75 milligrams for women.

Chapter Glossary

_____ 1. biotin

_____ 2. vitamin B_6

_____ 3. rickets

_____ 4. erythrocyte hemolysis

_____ 5. choline

_____ 6. pellagra

_____ 7. scurvy

_____ 8. night blindness

_____ 9. prooxidant

_____ 10. niacin

_____ 11. riboflavin

_____ 12. vitamin B_{12}

_____ 13. inositol

_____ 14. ascorbic acid

_____ 15. beriberi

_____ 16. pernicious anemia

_____ 17. folate

_____ 18. thiamin

_____ 19. carotenoids

_____ 20. osteomalacia

a. the vitamin C-deficiency disease

b. a nonessential nutrient found in cell membranes

c. a B vitamin needed in protein metabolism; its three active forms are pyridoxine, pyridoxal, and pyridoxamine

d. a B vitamin needed in energy metabolism; it can be eaten preformed or can be made in the body from tryptophan

e. a B vitamin that acts as part of a coenzyme important in the manufacture of new cells

f. rupture of the red blood cells, caused by vitamin E deficiency

g. the thiamin-deficiency disease

h. a B vitamin; a coenzyme necessary for fat synthesis and other metabolic reactions

i. one of the active forms of vitamin C

j. the niacin-deficiency disease

k. a B vitamin active in the body's energy-release mechanisms

l. the vitamin D-deficiency disease in children

m. slow recovery of vision after exposure to flashes of bright light at night; an early symptom of vitamin A deficiency

n. a B vitamin that helps convert folate to its active form and also helps maintain the sheaths around nerve cells

o. a compound that triggers reactions involving oxygen

p. a nonessential nutrient used to make the phospholipid lecithin and other molecules

q. a vitamin B_{12}-deficiency disease, caused by lack of intrinsic factor and characterized by large, immature red blood cells and damage to the nervous system

r. a B vitamin involved in the body's use of fuels which occupies a special site on the nerve cell membranes

s. the vitamin D deficiency disease in adults

t. a group of pigments in foods ranging from light yellow to reddish orange, many with vitamin A activity in the body

Crossword Puzzle:

Word Bank:			
beta-carotene	intrinsic factor	RAE	tocopherol
coenzyme	IU	retina	vitamins
collagen	keratin	retinol	xerophthalmia
cornea	precursors	rhodopsin	xerosis

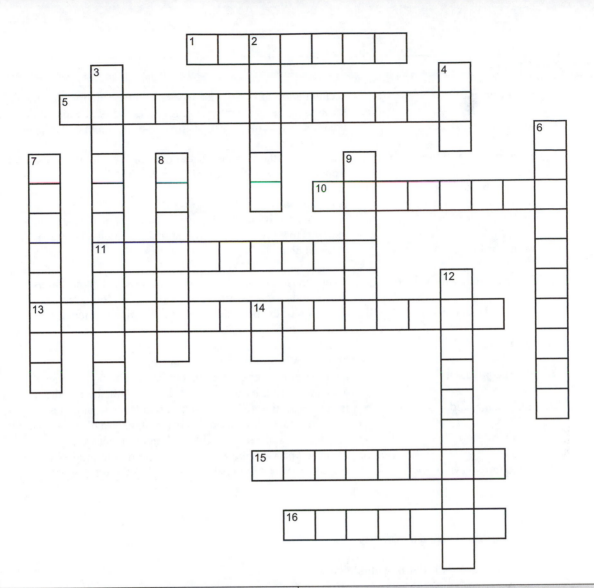

Across:

1. the normal protein of hair and nails
5. hardening of the cornea of the eye in advanced vitamin A deficiency that can lead to blindness
10. a small molecule that works with an enzyme to promote the enzyme's activity
11. the light-sensitive pigment of the cells in the retina
13. a factor found inside the digestive system and made by the stomach wall lining
15. the chief protein of most connective tissues, including scars, ligaments, and tendons, and the underlying matrix on which bones and teeth are built
16. one of the active forms of vitamin A made from beta-carotene in animal and human bodies

Down:

2. the layer of light-sensitive nerve cells lining the back of the inside of the eye
3. an orange pigment with antioxidant activity; a vitamin A precursor made by plants and stored in human fat tissue
4. a new measure of vitamin A activity of beta-carotene and other vitamin A precursors
6. compounds that can be converted into active vitamins
7. organic compounds that are vital to life and indispensable to body functions, but are needed only in minute amounts
8. drying of the cornea; a symptom of vitamin A deficiency
9. the hard, transparent membrane covering the outside of the eye
12. a kind of alcohol
14. a measure of fat-soluble vitamin activity sometimes used on supplement labels

Matching Exercise 2:

_____	1. homocysteine	a.	accumulation of keratin in a tissue; a sign of vitamin A deficiency
_____	2. dietary antioxidants	b.	the amount of niacin present in food, including the niacin that can theoretically be made from its precursor tryptophan that is present in the food
_____	3. carnitine	c.	yellowing of the skin due to spillover of the bile pigment bilirubin from the liver into the general circulation
_____	4. neural tube defects	d.	an amino acid produced as an intermediate compound during amino acid metabolism
_____	5. epithelial tissue	e.	abnormalities of the brain and spinal cord apparent at birth and believed to be related to a woman's folate intake before and during pregnancy
_____	6. keratinization	f.	a B vitamin
_____	7. Dietary Folate Equivalent (DFE)	g.	compounds typically found in plant foods that significantly decrease the adverse effects of oxidation on living tissues
_____	8. macular degeneration	h.	the layers of the body that serve as selective barriers to environmental factors
_____	9. free radicals	i.	a unit of measure expressing the amount of folate available to the body from naturally occurring sources
_____	10. niacin equivalents	j.	a nonessential nutrient that functions in cellular activities
_____	11. pantothenic acid	k.	a common, progressive loss of function of the part of the retina that is most crucial to focused vision; often leads to blindness
_____	12. jaundice	l.	atoms or molecules with one or more unpaired electrons that make the atom or molecule unstable and highly reactive
_____	13. carpal tunnel syndrome	m.	a pinched nerve at the wrist, causing pain or numbness in the hand; it is often caused by repetitive motion of the wrist

Exercises

Answer these chapter study questions:

1. List the general characteristics of fat-soluble vitamins.

2. Do individuals have to eat vitamin D to have enough in their bodies? Why or why not?

3. Name three reasons why it is almost impossible to create vitamin E deficiencies in humans.

4. Why would someone who had part of the stomach removed develop deficiency symptoms for vitamin B_{12}?

5. Is it possible for a person consuming adequate protein and adequate calories to be deficient in niacin? Why or why not?

6. What has been responsible for the almost 20 percent drop in the national incidence of neural tube defects?

Complete these short answer questions:

1. The fat soluble vitamins include:

    a.
    b.
    c.
    d.

2. Vitamin A has parts to play in such diverse functions as:

    a.
    b.
    c.
    d.
    e.
    f.

3. The richest food sources of beta carotene include:

    a.
    b.
    c.
    d.
    e.
    f.
    g.
    h.

4. The earliest symptoms of vitamin A overdoses are:

    a.
    b.
    c.
    d.
    e.
    f.

5. To raise the level of blood calcium, the body can draw from these places:

    a.
    b.
    c.

6. Vitamin K's richest plant sources are:

    a.
    b.

7. These individuals may incur, over a period of years, a vitamin D deficiency:

    a.
    b.
    c.

8. In adults, vitamin E deficiency is usually associated with diseases of these organs:

    a.
    b.
    c.

9. The best natural sources of folate are _____ vegetables and fruits.

10. In the U.S. today, scurvy may be seen in these groups:

    a.
    b.
    c.

Solve these problems:

1. Your friend is taking supplements of the antioxidant nutrients in an effort to reduce her risk of cancer. Based on the latest research in this area, what advice would you offer your friend?

2. If a person consumed a diet composed of 1200 micrograms of beta-carotene from food, how many micrograms of retinol would this supply in the body?

Answer these controversy questions:

1. What are some of the general labeling requirements for supplements based on the Dietary Supplement Health and Education Act (DSHEA)?

2. What are some of the problems that have been identified when supplements have been tested?

3. What is the primary characteristic of the diet of populations with low cancer rates?

4. What is a potential harmful effect of taking vitamin supplements?

5. You see a USP symbol on the label of a supplement. What does this symbol tell the consumer?

Study Aids

1. Complete the following chart which identifies names, chief functions, deficiency disease names, and major food sources of some of the fat-soluble and water-soluble vitamins.

Name	Chief Functions	Deficiency Disease Name(s)	Food Sources
A	<ul><li>_____</li><li>bone _____</li><li>reproduction</li><li>_____ cells</li></ul>	_____	*For retinol:* • fortified _____ or _____ • cheese • _____ or cream • eggs • _____ *For beta-carotene:* • dark leafy _____ • deep orange fruits and _____

Name	Chief Functions	Deficiency Disease Name(s)	Food Sources
D	• mineralization of _____	rickets; _____	• _____ milk • eggs • liver • _____
E	• _____ • protects PUFA and vitamin _____	no name	• polyunsaturated plant _____ • _____ germ • green leafy vegetables
K	• synthesis of blood _____ proteins	no name	• _____ • _____-type vegetables • green _____ vegetables • _____ oils
Thiamin	• part of a _____ used in energy metabolism	_____	• occurs in all nutritious foods
Riboflavin	• part of a coenzyme used in _____ metabolism	_____	• milk, yogurt and cottage cheese • meat and _____ • leafy green vegetables • whole grain or _____ breads and _____
Niacin	• part of a _____ used in energy metabolism	_____	• milk • _____ • meat • _____ • fish • enriched _____ and cereals • all _____- containing foods
Vitamin C	• _____ synthesis • _____ • hormone synthesis • _____ acid metabolism • strengthens resistance to _____ • helps in absorption of _____	_____	• _____ fruits • _____- type vegetables • cantaloupe • _____ • peppers • lettuce • _____ • potatoes • _____ or mangoes

2. Use Table C7-1 on page 255 of your textbook to study valid reasons for taking supplements.

Sample Test Items

Comprehension Level Items:

1. The vitamins fall naturally into _____ classes.

    a. two
    b. three
    c. four
    d. five

2. All of the following are characteristics of fat-soluble vitamins **except**:

    a. they require bile for absorption.
    b. they can reach toxic levels.
    c. they are easily excreted.
    d. they are stored in fatty tissues.

3. The first fat-soluble vitamin to be recognized was vitamin:

    a. D.
    b. K.
    c. E.
    d. A.

4. A vitamin A precursor found in plants is:

    a. called beta carotene.
    b. a bright orange color.
    c. yellow in color.
    d. a and b
    e. a and c

5. All of the following are good sources of beta carotene **except**:

    a. carrots.
    b. pumpkins.
    c. red cabbage.
    d. apricots.

6. The vitamin D deficiency disease in children is known as:

    a. osteodystrophy.
    b. rickets.
    c. pellagra.
    d. osteomalacia.

7. People who are exposed to the sun make vitamin D from:

    a. cholesterol.
    b. carotene.
    c. polyunsaturated fats.
    d. saturated fats.

8. Vitamin E exerts an especially important antioxidant effect in the:

    a. kidneys.
    b. heart.
    c. lungs.
    d. stomach.

9. The DRI for vitamin E is _____ milligrams a day for adults.

    a. 5
    b. 8
    c. 10
    d. 15

10. Vitamin K deficiencies may occur in:

    a. newborn infants.
    b. people who have undergone surgery.
    c. people who have taken antibiotics.
    d. a and c
    e. b and c

11. The niacin deficiency disease is known as:

    a. osteomalacia.
    b. scurvy.
    c. pellagra.
    d. beriberi.

12. A high intake of _____ can mask the anemia caused by vitamin B_{12} deficiency.

    a. thiamin
    b. folate
    c. riboflavin
    d. niacin

13. What is the amount of vitamin C required to prevent the symptoms of scurvy from appearing?

    a. 10 milligrams
    b. 20 milligrams
    c. 30 milligrams
    d. 40 milligrams

14. The amount needed of this vitamin is proportional to protein intake:

    a. thiamin.
    b. riboflavin.
    c. niacin.
    d. vitamin B_6.

15. People addicted to alcohol may develop a deficiency of:

    a. niacin.
    b. thiamin.
    c. vitamin B_6.
    d. vitamin B_{12}.

16. Vitamin B_{12} is naturally occurring only in:

    a. grains.
    b. plants.
    c. animals.
    d. vegetables.

17. In the U.S. today, scurvy has been found in all of the following **except**:

    a. breast-fed infants.
    b. infants fed only cow's milk.
    c. elderly.
    d. people addicted to drugs.

18. Which of the following supplements would be the most appropriate to choose?

    a. one that comes in a chewable form
    b. one that provides all the vitamins and minerals in amounts smaller than or close to the intake recommendations
    c. one that provides more than the DRI recommended intake for vitamin A, vitamin D and minerals
    d. one that provides more than the Tolerable Upper Intake Level for any nutrient

19. Vitamin C supplements in any dosage may be dangerous for people with an overload of _____ in the blood.

    a. zinc
    b. calcium
    c. magnesium
    d. iron

20. The adult DRI intake recommendation for vitamin C is _____ milligrams for women.

    a. 75
    b. 90
    c. 105
    d. 125

21. Which of the following has gained a reputation as an "anti-infective" vitamin?

    a. D
    b. A
    c. E
    d. K

110

22. Which of the following food sources of vitamin A should **not** be eaten on a daily basis because of concerns about toxicity?

    a. fish oil
    b. eggs
    c. liver
    d. butter

23. The most potentially toxic of all vitamins is vitamin:

    a. A.
    b. C.
    c. D.
    d. K.

24. Diseases of the liver, gallbladder, and pancreas are likely to result in a deficiency of vitamin

    a. A.
    b. C.
    c. K.
    d. E.

Application Level Items:

25. A friend of yours complains that she is unable to see for a short period of time after she encounters a flash of bright light at night. Which of the following statements would you make to your friend?

    a. Your eyes are probably well nourished and normal.
    b. You are suffering from vitamin A deficiency.
    c. You have night blindness.
    d. You may need to check your diet for vitamin A deficiency.

26. A couple you know gives their infant a jar of baby carrots to eat every day. This child runs the risk of:

    a. developing an allergic reaction to carrots.
    b. developing a vitamin A toxicity.
    c. having his skin turn bright yellow.
    d. a and b
    e. b and c

27. Which of the following would have the lowest need for vitamin A?

    a. a man
    b. a child
    c. a woman
    d. a lactating woman

28. You have a friend who has been experiencing appetite loss, nausea, vomiting, and increased urination and thirst. A medical examination reveals that calcium has been deposited in the heart and kidneys. Based on these symptoms you suspect that your friend has been taking supplements of:

    a. vitamin A.
    b. vitamin K.
    c. vitamin D.
    d. vitamin E.

29. Which of the following would **not** be at risk for developing a vitamin D deficiency?

    a. a woman who is institutionalized
    b. a man who works the night shift for several years
    c. a dark-skinned child who lives in a northern city
    d. an adult who eats cereal with milk for breakfast every day

30. An athlete who consumes a high-protein diet would have an increased need for:

    a. vitamin B_6.
    b. vitamin C.
    c. vitamin B_{12}.
    d. thiamin.

Answers

Summing Up

(1) tiny; (2) absorbed; (3) structures; (4) precursors; (5) fat; (6) water; (7) vision; (8) retinol; (9) beta carotene; (10) cantaloupe: (11) bone; (12) children; (13) osteomalacia; (14) toxic; (15) cholesterol; (16) 5; (17) antioxidant; (18) lungs; (19) fat; (20) 15; (21) vegetable; (22) blood clotting; (23) intestinal; (24) antibiotics; (25) water-soluble; (26) coenzymes; (27) energy; (28) vitamin B_6; (29) protein; (30) beriberi; (31) niacin; (32) tryptophan; (33) vitamin B_{12}; (34) 10; (35) 90

Chapter Glossary

Matching Exercise 1: (1) h; (2) c; (3) l; (4) f; (5) p; (6) j; (7) a; (8) m; (9) o; (10) d; (11) k; (12) n; (13) b; (14) i; (15) g; (16) q; (17) e; (18) r; (19) t; (20) s
Crossword Puzzle: (1) keratin; (2) retina; (3) beta-carotene; (4) RAE; (5) xerophthalmia; (6) precursors; (7) vitamins; (8) xerosis; (9) cornea; (10) coenzyme; (11) rhodopsin; (12) tocopherol; (13) intrinsic factor; (14) IU; (15) collagen; (16) retinol
Matching Exercise 2: (1) d; (2) g; (3) j; (4) e; (5) h; (6) a; (7) i; (8) k; (9) l; (10) b; (11) f; (12) c; (13) m

Exercises

Chapter Study Questions:

1. They require bile for absorption. They generally occur together in the fats and oils of foods; once they have been absorbed from the intestinal tract, they are stored in the liver and fatty tissues until the body needs them; excesses can reach toxic levels; they can be lost from the digestive tract with undigested fat.

2. Humans do not have to eat vitamin D because they can make it from a cholesterol compound whenever ultraviolet light from the sun shines on their skin. The compound is transformed into a vitamin D precursor and is absorbed directly into the blood.

3. Because vitamin E is so widespread in food it is almost impossible to create a vitamin E deficient diet. In addition, the body stores so much vitamin E in its fatty tissues that a person could not keep on eating a vitamin E free diet for long enough to deplete those stores and produce a deficiency. Finally, the cells recycle their working supply of vitamin E, using the same molecules over and over.

4. The absorption of vitamin B_{12} requires an intrinsic factor, which is synthesized in the stomach. When part of the stomach has been removed, it cannot produce enough of the intrinsic factor. Without enough of the intrinsic factor, the person cannot absorb the vitamin even though he may be getting enough in the diet.

5. A person eating adequate protein will not be deficient in niacin because the amino acid tryptophan is converted to niacin in the body.

6. In the 1990s, the FDA ordered fortification of all enriched grain products with folic acid, which has resulted in dramatic increases in folate intakes. Folate deficiency is associated with neural tube defects.

Short Answer Questions:

1. (a) A; (b) D; (c) E; (d) K

2. (a) vision; (b) maintenance of body linings and skin; (c) bone and body growth; (d) reproduction; (e) normal cell development; (f) immune defenses

3. (a) carrots; (b) sweet potatoes; (c) pumpkins; (d) cantaloupe; (e) apricots; (f) spinach; (g) broccoli; (h) mango

4. (a) blurred vision; (b) growth failure in children; (c) loss of appetite; (d) headache; (e) itching of skin; (f) irritability

5. (a) bones; (b) food in the digestive tract; (c) kidneys, which recycle calcium

6. (a) dark-green leafy vegetables; (b) members of the cabbage family

7. (a) housebound; (b) institutionalized; (c) those who work at night

8. (a) liver; (b) gallbladder; (c) pancreas

9. uncooked

10. (a) infants fed only cow's milk; (b) elderly; (c) people addicted to alcohol or other drugs

Problem-Solving:

1. Research findings thus far do not support taking antioxidant supplements to prevent cancer. More research is needed to quantify them and to clarify the risks and benefits. However, there is strong support for recommendations to increase consumption of fruits and vegetables to provide these nutrients.

2. 100 micrograms of retinol

Controversy Questions:

1. They must provide nutrient information; they must not be adulterated; they may not claim to diagnose, prevent, cure or treat specific illnesses; they may bear structure/function claims.

2. Supplements have been found to contain substances not listed on their labels, to lack the listed substances, and to vary in potency or variety from the label information.

3. They consume abundant vegetables and fruits, especially those containing antioxidants, such as beta-carotene and other carotenoids.

4. Vitamin C supplements enhance iron absorption, making iron overload likely in some people.

5. The USP symbol means that a manufacturer has voluntarily paid an independent laboratory to test the product and affirm that it contains the ingredients as listed and that it will dissolve or disintegrate in the digestive tract to make the ingredients available for absorption.

Study Aids

Name	Chief Functions	Deficiency Disease Name(s)	Food Sources
A	• vision • bone growth • reproduction • epithelial cells	hypovitaminosis A	*For retinol:* • fortified milk or margarine • cheese • butter or cream • eggs • liver *For beta-carotene:* • dark leafy greens • deep orange fruits and vegetables
D	• mineralization of bones	rickets; osteomalacia	• fortified milk • eggs • liver • sardines
E	• antioxidant • protects PUFA and vitamin A	no name	• polyunsaturated plant oils • wheat germ • green, leafy vegetables
K	• synthesis of blood clotting proteins	no name	• soybeans • cabbage-type vegetables • green leafy vegetables • vegetable oils
Thiamin	• part of a coenzyme used in energy metabolism	beriberi	• occurs in all nutritious foods
Riboflavin	• part of a coenzyme used in energy metabolism	ariboflavinosis	• milk, yogurt and cottage cheese • meat and liver • leafy green vegetables • whole grain or enriched breads and cereals
Niacin	• part of a coenzyme used in energy metabolism	pellagra	• milk • eggs • meat • poultry • fish • enriched breads and cereals • all protein-containing foods

Name	Chief Functions	Deficiency Disease Name(s)	Food Sources
Vitamin C	<ul><li>collagen synthesis</li><li>antioxidant</li><li>hormone synthesis</li><li>amino acid metabolism</li><li>strengthens resistance to infection</li><li>helps in absorption of iron</li></ul>	scurvy	<ul><li>citrus fruits</li><li>cabbage-type vegetables</li><li>cantaloupe</li><li>strawberries</li><li>peppers</li><li>lettuce</li><li>tomatoes</li><li>potatoes</li><li>papayas or mangoes</li></ul>

Sample Test Items

1. a (p. 214)
2. c (p. 215)
3. d (p. 215)
4. d (p. 215)
5. c (p. 220)
6. b (p. 221)
7. a (p. 222)
8. c (p. 224)

9. d (p. 225)
10. d (p. 227)
11. c (p. 236)
12. b (p. 239, 241)
13. a (p. 231)
14. d (p. 242)
15. b (p. 235)
16. c (p. 241)

17. a (p. 230)
18. b (p. 261)
19. d (p. 230)
20. a (p. 231)
21. b (p. 217)
22. c (p. 218)
23. c (p. 222)
24. d (p. 225)

25. a (p. 216)
26. c (p. 220)
27. b (p. 219)
28. c (p. 222)
29. d (p. 222-223)
30. a (p. 242)

Chapter 8 - Water and Minerals

Chapter Objectives

After completing this chapter, you should be able to:

1. Discuss the major roles of water in the body and the amount of water needed by adults.

2. Describe how the body regulates water intake and excretion to maintain water balance.

3. Compare the arguments for and against drinking bottled water.

4. Explain the role of minerals in maintaining the body's fluid and electrolyte balance and acid-base balance.

5. List the major roles and important deficiency and toxicity symptoms for each major and trace mineral.

6. List good food sources for the major and trace elements.

7. Discuss the lifestyle choices that reduce the risk of osteoporosis (Controversy 8).

Key Concepts

✓ Water provides the medium for transportation, acts as a solvent, participates in chemical reactions, provides lubrication and shock protection, and aids in temperature regulation in the human body.

✓ Water makes up about 60 percent of the body's weight. A change in the body's water content can bring about a temporary change in body weight.

✓ Water losses from the body necessitate intake equal to output to maintain balance. The brain regulates water intake; the brain and kidneys regulate water excretion. Dehydration and water intoxication can have serious consequences.

✓ Many factors influence a person's need for water. The water of beverages and foods meets nearly all of the need for water, and a little more is supplied by the water formed during cellular breakdown of energy nutrients.

✓ Hard water is high in calcium and magnesium. Soft water is high in sodium, and it dissolves cadmium and lead from pipes.

✓ Public drinking water is tested and treated for safety. All drinking water originates from surface water or ground water that is vulnerable to contamination from human activities.

✓ Mineral salts form electrolytes that help keep fluids in their proper compartments and buffer these fluids, permitting all life processes to take place.

✓ Calcium makes up bone and tooth structure and plays roles in nerve transmission, muscle contraction, and blood clotting. Calcium absorption rises when there is a dietary deficiency or an increased need such as during growth.

- Most of the phosphorus in the body is in the bones and teeth. Phosphorus helps maintain acid-base balance, is part of the genetic material in cells, assists in energy metabolism, and forms part of cell membranes. Under normal circumstances, deficiencies of phosphorus are unknown.

- Most of the body's magnesium is in the bones and can be drawn out for all the cells to use in building protein and using energy. Most people in the United States fail to obtain enough magnesium from their food.

- Sodium is the main positively charged ion outside the body's cells. Sodium attracts water. Thus, too much sodium (or salt) raises blood pressure and aggravates hypertension. Diets rarely lack sodium.

- Potassium, the major positive ion inside cells, is important in many metabolic functions. Fresh whole foods are the best sources of potassium. Diuretics can deplete the body's potassium and so can be dangerous; potassium excess can also be dangerous.

- Chloride is the body's major negative ion; it is responsible for stomach acidity and assists in maintaining proper body chemistry. No known diet lacks chloride.

- Sulfate is a necessary nutrient used to synthesize sulfur-containing body compounds.

- Iodine is part of the hormone thyroxine, which influences energy metabolism. The deficiency diseases are goiter and cretinism. Iodine occurs naturally in seafood and in foods grown on land that was once covered by oceans; it is an additive in milk and bakery products. Large amounts are poisonous. Potassium iodide, appropriately administered, blocks some radiation damage to the thyroid during radiation emergencies.

- Most iron in the body is contained in hemoglobin and myoglobin or occurs as part of enzymes in the energy-yielding pathways. Iron-deficiency anemia is a problem worldwide; too much iron is toxic. Iron is lost through menstruation and other bleeding; reduced absorption and the shedding of intestinal cells protect against overload. For maximum iron absorption, use meat, other iron sources, and vitamin C together.

- Zinc assists enzymes in all cells. Deficiencies in children cause growth retardation with sexual immaturity. Zinc supplements can reach toxic doses, but zinc in foods is nontoxic. Animal foods are the best sources.

- Selenium works with an enzyme system to protect body compounds from oxidation. A deficiency induces a disease of the heart. Deficiencies are rare in developed countries, but toxicities can occur from overuse of supplements.

- Fluoride stabilizes bones and makes teeth resistant to decay. Excess fluoride discolors teeth; large doses are toxic.

- Chromium works with the hormone insulin to control blood glucose concentrations. Chromium is present in a variety of unrefined foods.

- Copper is needed to form hemoglobin and collagen and assists in many other body processes. Copper deficiency is rare.

- Many different trace elements play important roles in the body. All of the trace minerals are toxic in excess.

Summing Up

Water makes up about (1)_____ percent of all the body's weight and is the most indispensable

nutrient of all. Thirst and (2)_____ govern water intake, although (3)_____ lags

behind a lack of water. Water excretion is governed by the brain and the (4)_____, while the

(5)_____ regulates water intake. A change in the body's water content can bring a change in body

(6)_____. Water losses from the body necessitate intake of an equal amount of water to maintain

water (7)_____. In the body, water provides the medium for (8)_____, chemical

reactions, shock protection, lubrication, and (9)_____ regulation.

Under normal dietary and environmental conditions, men need about (10)_____ cups of fluid

from beverages and drinking water and women need about 9 cups. This amount of fluid provides about

(11)_____ percent of the day's need for water. Hard water is high in calcium and

(12)_____, while soft water is high in (13)_____.

Most of the body's water weight is contained inside the (14)_____, and some water also

bathes the (15)_____ of the cells. Body cells can pump (16)_____ across their

membranes, but not water, and these minerals attract the water to come along with them. The cells use minerals for

this purpose in a special form, known as ions or (17)_____. The result of the system's working

properly is (18)_____ and electrolyte balance. The minerals also help manage another balancing

act, the (19)_____ balance. Among the major minerals, some, when dissolved in water, give rise to

acids, and some to (20)_____.

Calcium is by far the most (21)_____ mineral in the body. Calcium makes up bone and

(22)_____ structure and plays roles in (23)_____ transmission, muscle contraction and

blood clotting. Adult bone loss is called (24)_____. The DRI recommended intake for calcium has been

set at (25)_____ milligrams for adolescents, because people develop their peak bone mass during their

growing years.

Sodium is the chief ion used to maintain the volume of fluid (26)_____ cells and it contributes

(27)_____ percent of the weight of the compound sodium chloride. Sodium helps maintain

(28)_____ balance and is essential to nerve transmission and muscle contraction. The DRI intake

recommendation for sodium has been set at (29)_____ milligrams for healthy, active young adults.

Both potassium deficiency and excess can be dangerous and (30)_____ can deplete body

potassium. Other major minerals include (31)_____, which helps strands of protein to assume and

hold a functional shape, magnesium, phosphorus and (32)_____, which is part of the stomach's

hydrochloric acid.

One of the trace minerals is iodine, which is part of the hormone (33)_____, which regulates the basal metabolic rate. Every living cell contains (34)_____, which is contained in hemoglobin in red blood cells and (35)_____ in muscle cells. Zinc works with proteins in every organ as a helper for nearly 100 (36)_____ and the best sources are animal foods. Fluoride stabilizes bone and makes (37)_____ resistant to decay.

Chapter Glossary

Matching Exercise:

_____ 1. hemoglobin

_____ 2. osteoporosis

_____ 3. fluid and electrolyte balance

_____ 4. goiter

_____ 5. phytates

_____ 6. peak bone mass

_____ 7. iron overload

_____ 8. buffers

_____ 9. tannins

_____ 10. heme

_____ 11. dehydration

_____ 12. iron deficiency

_____ 13. hydroxyapatite

_____ 14. cretinism

_____ 15. MFP factor

_____ 16. acid-base balance

_____ 17. myoglobin

_____ 18. pica

a. compounds in tea (especially black tea), and coffee that bind iron

b. enlargement of the thyroid gland due to iodine deficiency

c. the oxygen-holding protein of the muscles

d. the highest attainable bone density for an individual

e. the oxygen-carrying protein of the blood

f. severe mental and physical retardation of an infant caused by the mother's iodine deficiency during her pregnancy

g. maintenance of the proper amounts and kinds of fluids and minerals in each compartment of the body

h. the condition of having depleted iron stores, which, at the extreme, causes iron-deficiency anemia

i. maintenance of the proper degree of acidity in each of the body's fluids

j. a reduction of the bone mass of older persons in which the bones become porous and fragile

k. molecules that can help to keep the pH of a solution from changing by gathering or releasing H ions

l. the chief crystal of bone, formed from calcium and phosphorus

m. loss of water

n. a craving for nonfood substances

o. the state of having more iron in the body than it needs or can handle, usually from a hereditary defect

p. a factor (identity unknown) present in meat, fish, and poultry that enhances the absorption of nonheme iron present in the same foods or in other foods eaten at the same time

q. the iron-containing portion of the hemoglobin and myoglobin molecules

r. compounds present in plant foods (particularly whole grains) that bind iron and prevent its absorption

Crossword Puzzle:

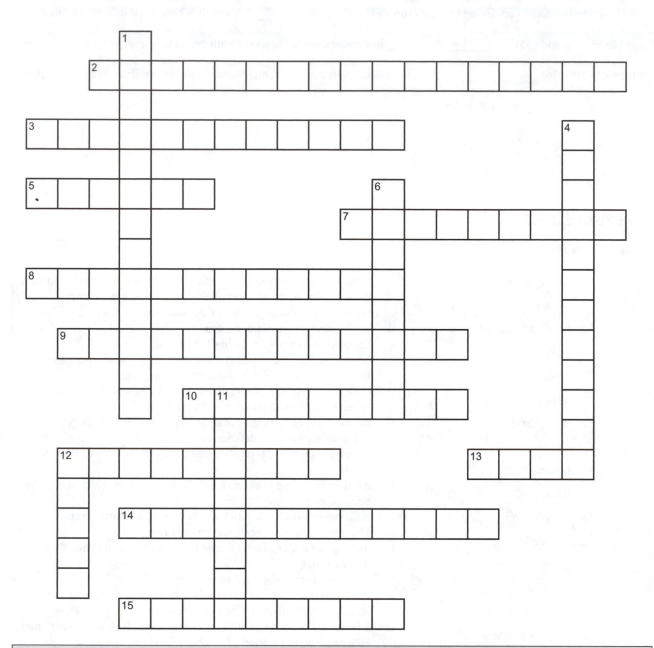

Word Bank:

anemia

diuretics

electrolytes

fluorapatite

fluorosis

hard water

hypertension

ions

leavened

major minerals

minerals

salts

soft water

trace minerals

water balance

water intoxication

Crossword Puzzle Across:	Crossword Puzzle Down:
2. a dangerous dilution of the body's fluids, usually from excessive ingestion of plain water	1. essential mineral nutrients found in the human body in amounts larger than 5 grams
3. a crystal of bones and teeth, formed when fluoride displaces the hydroxy portion of hydroxyapatite	4. compounds that partly dissociate in water to form ions, such as the potassium ion (K+) and the chloride ion (Cl-)
5. the condition of inadequate or impaired red blood cells	6. naturally occurring, inorganic, homogenous substances; chemical elements
7. compounds, usually medications, causing increased urinary water excretion	11. literally, "*lightened*" by yeast cells, which digest some carbohydrate components of the dough and leave behind bubbles of gas that make the bread rise
8. the balance between water intake and water excretion, which keeps the body's water content constant	12. compounds composed of charged particles
9. essential mineral nutrients found in the human body in amounts less than 5 grams	
10. discoloration of the teeth due to ingestion of too much fluoride during tooth development	
12. water with a high sodium concentration	
13. electrically charged particles, such as sodium (positively charged) or chloride (negatively charged)	
14. high blood pressure	
15. water with high calcium and magnesium concentrations	

Exercises

Answer these chapter study questions:

1. Discuss the health implications of drinking hard water versus soft water.

2. Why is setting recommended intakes for calcium difficult?

3. Why are recommended intakes for calcium higher for children and adolescents?

4. Why should someone taking diuretics, which cause potassium loss, be advised to eat potassium-rich foods?

5. Define the term *pica*. What type of mineral deficiency is associated with pica?

6. What does DASH stand for, and what is are the characteristics of the DASH diet?

Complete these short answer questions:

1. The major minerals include:

   a.
   b.
   c.
   d.
   e.
   f.
   g.

2. The distinction between hard water and soft water is based on these three minerals:

    a.
    b.
    c.

3. The major roles of calcium in the body's fluids include:

    a.

    b.

    c.

    d.

    e.

    f.

4. All of the following can deplete the body of needed sodium:

    a.
    b.
    c.
    d.

5. The person who wishes to learn to control salt intake should:

    a.
    b.

6. Deficiency of magnesium may occur as a result of:

    a.
    b.
    c.
    d.
    e.

7. The need for iodine is easily met by consuming:

    a.
    b.

8. Factors that hinder iron absorption include:

    a.
    b.
    c.

Solve these problems:

Modify the following recipe to make it lower in calories, fat, and sodium and to substitute polyunsaturated fat for saturated fat.

Chicken Pot Pie

Pastry
1 cup flour
¼ tsp. salt
2 tbsp. water
$^{1}/_{3}$ cup lard

Filling
$^{1}/_{3}$ cup chopped potatoes
$^{1}/_{3}$ cup sliced carrots
$^{1}/_{3}$ cup canned green peas
¼ cup chopped celery
1 tbsp. chopped onion
½ cup boiling water
¼ cup butter
¼ cup flour
½ tsp. salt
$^{1}/_{8}$ tsp. pepper
$^{1}/_{8}$ tsp. poultry seasoning
1 $^{1}/_{3}$ cup chicken broth
$^{2}/_{3}$ cup cream
1 ½ cup diced, cooked chicken

Modifications:

Answer these controversy questions:

1. Differentiate between trabecular bone and cortical bone.

2. What are environmental factors currently under study for their roles in lowering bone density?

3. How is genetics thought to influence bone density?

4. What types of physical activity are recommended for building muscle and bone strength to reduce the risk of osteoporosis?

5. What are the guidelines for growing healthy bones during childhood?

Study Aids

1. Complete the following chart which identifies the names, chief functions, deficiency disease names, and major food sources of some of the major and trace minerals.

Major Minerals			
Name	**Chief Functions**	**Deficiency Disease Name**	**Food Sources**
Calcium	• makes up bone and _____ structure • normal muscle contraction • nerve functioning • blood _____ • immune defenses • blood pressure	_____	• milk • small _____ with bones • tofu • greens • legumes
Magnesium	• _____ mineralization • building of protein • normal muscle contraction • transmission of nerve _____ • maintenance of _____		• nuts • _____ • whole grains • dark green _____ • seafoods • _____ • cocoa
Sodium	• maintains normal fluid balance • maintains normal _____ balance		• salt • _____ foods • _____ sauce

Trace Minerals			
Name	**Chief Functions**	**Deficiency Disease Name**	**Food Sources**
Iron	• part of the protein _____ • part of the protein myoglobin • necessary for utilization of _____ in the body	_____	• _____ meats • _____ • poultry • shellfish • eggs • legumes • _____ fruits
Iodine	• component of thyroid hormone _____	_____ _____	• _____ • _____ salt • bread
Zinc	• part of insulin and many _____ • normal _____ development		• _____- containing foods

2. Identify the intake recommendations for calcium, sodium, and iron by studying Tables C8-3, 8-5, and 8-9, respectively, on pages 309, 279, and 290 of your textbook.

Sample Test Items

Comprehension Level Items:

1. The distinction between the major and trace minerals is that:

    a. the major minerals are more important than the trace minerals.
    b. the major minerals play more important roles than the trace minerals.
    c. the major minerals are present in larger quantities than the trace minerals.
    d. a and c
    e. b and c

2. All of the following play a role in water intake **except**:

    a. the mouth.
    b. the volume of blood.
    c. the hypothalamus.
    d. the lungs.

3. One of the principal minerals in hard water is:

    a. calcium.
    b. fluoride.
    c. sodium.
    d. potassium.

4. Water needs vary greatly depending on:

    a. the food a person eats.
    b. the environmental temperature and humidity.
    c. the person's gender.
    d. a and b
    e. b and c

5. The body's proteins and some of its minerals prevent changes in the acid-base balance of its fluids by serving as:

    a. ions.
    b. electrolytes.
    c. buffers.
    d. solutes.

6. Adult bone loss is referred to as:

    a. rickets.
    b. osteomalacia.
    c. osteodystrophy.
    d. osteoporosis.

7. Which of the following is(are) characteristics of osteoporosis?

   a. bones become brittle
   b. bones become fragile
   c. bones bend
   d. a and b
   e. b and c

8. Which of the following groups absorbs the greatest amount of calcium?

   a. infants
   b. pregnant women
   c. adolescents
   d. older adults

9. Which of the following is the best source of calcium?

   a. butter
   b. cream
   c. cream cheese
   d. yogurt

10. The DRI Committee has set a Tolerable Upper Intake Level for sodium of _____ milligrams per day.

    a. 300
    b. 500
    c. 2300
    d. 3200

11. Processed and fast foods are the source of almost _____ percent of the salt in most people's diets.

    a. 15
    b. 35
    c. 52
    d. 75

12. The most productive step to take to control salt intake is to:

    a. stop using instant foods.
    b. discontinue the use of milk products.
    c. limit intake of processed and fast foods.
    d. control use of the salt shaker.

13. Which of the following can cause potassium depletion?

    a. use of diuretics that cause potassium loss
    b. severe diarrhea
    c. excessive use of potassium chloride
    d. a and b
    e. b and c

14. Which of the following is part of the hydrochloric acid that maintains the strong acidity of the stomach?

   a. chloride
   b. sulfur
   c. magnesium
   d. potassium

15. Foods rich in magnesium are:

   a. slightly processed.
   b. unprocessed.
   c. highly processed.
   d. a and c
   e. a and b

16. Which of the following is **not** a completely dependable source of iodine?

   a. seafood
   b. sea salt
   c. iodized salt
   d. plants grown in iodine-rich soil

17. All of the following are symptoms of iron deficiency **except**:

   a. tiredness.
   b. impaired concentration.
   c. intolerance to heat.
   d. apathy.

18. Which of the following is often responsible for low iron intakes in the western world?

   a. blood loss caused by parasitic infections of the digestive tract
   b. malnutrition which includes an inadequate iron intake
   c. vegetarians who design their own meal patterns
   d. displacement of nutrient-rich foods by foods high in sugar and fat

19. The usual source of fluoride is:

   a. drinking water.
   b. cooking utensils.
   c. fortified foods.
   d. processed foods.

20. People with severe iodine deficiency:

   a. become sluggish.
   b. develop a goiter.
   c. lose weight.
   d. a and b
   e. b and c

21. Temporary fluctuations in body water

    a. reflect a change in body fat.
    b. show up on the scale.
    c. reflect gain or loss in water weight.
    d. a and b
    e. b and c

22. Given a normal diet and moderate environmental conditions, women need about _____ cups of fluid from drinking water and beverages per day.

    a. 3
    b. 5
    c. 9
    d. 13

23. Which of the following is the best source of phosphorus?

    a. steak
    b. beans
    c. bread
    d. cabbage

24. The DASH diet calls for greatly increased intakes of _____.

    a. nuts
    b. whole grains
    c. fish
    d. fruits and vegetables

Application Level Items:

25. Which of the following determines whether calcium is withdrawn from or deposited to the skeleton?

    a. hormones that are sensitive to blood levels of calcium
    b. the amount of calcium consumed in the diet
    c. the amount of calcium deposited in bones during peak bone mass
    d. a and b
    e. b and c

26. Why is the intake recommendation for calcium set at 1300 milligrams/day for adolescents?

    a. because the skeleton begins to lose bone density during that age
    b. because people develop their peak bone mass during this time
    c. because a high calcium intake is thought to maximize bone density during the growing years
    d. a and b
    e. b and c

27. Which of the following provides the best choices for high-calcium food selections?

    a. ice milk, cottage cheese, butter, and cream
    b. milk, buttermilk, cheese, and yogurt
    c. cream cheese, chocolate milk, and frozen yogurt
    d. sardines, spinach, kale, and Swiss chard

28. How much sodium is consumed by someone who eats five grams of salt?

    a. 1000 mg
    b. 1500 mg
    c. 2000 mg
    d. 2500 mg

29. An increase in iodine intake in the United States has resulted from:

    a. dough conditioners used in the baking industry.
    b. milk produced in dairies that feed cows iodine-containing medications.
    c. the increased consumption of fast foods and convenience items.
    d. a and b
    e. a and c

30. To improve the absorption of iron from an iron supplement you would:

    a. take the supplement with orange juice.
    b. take the supplement with tea.
    c. consume a small amount of meat along with the supplement.
    d. a and b
    e. a and c

Answers

Summing Up

(1) 60; (2) satiety; (3) thirst; (4) kidneys; (5) brain; (6) weight; (7) balance; (8) transportation; (9) temperature;
(10) 13; (11) 80; (12) magnesium, (13) sodium; (14) cells; (15) outsides; (16) minerals; (17) electrolytes; (18) fluid;
(19) acid-base; (20) bases; (21) abundant; (22) tooth; (23) nerve; (24) osteoporosis; (25) 1,300; (26) outside;
(27) 40; (28) acid-base; (29) 1500; (30) diuretics; (31) sulfate; (32) chloride; (33) thyroxine; (34) iron;
(35) myoglobin; (36) enzymes; (37) teeth

Chapter Glossary

Matching Exercise: (1) e; (2) j; (3) g; (4) b; (5) r; (6) d; (7) o; (8) k; (9) a; (10) q; (11) m; (12) h; (13) l; (14) f;
(15) p; (16) i; (17) c; (18) n
Crossword Puzzle: (1) major minerals; (2) water intoxication; (3) fluorapatite; (4) electrolytes; (5) anemia;
(6) minerals; (7) diuretics; (8) water balance; (9) trace minerals; (10) fluorosis; (11) leavened; (12 A) soft water;
(12 D) salts; (13) ions; (14) hypertension; (15) hard water

Exercises

Chapter Study Questions:

1. Hard water has high concentrations of calcium and magnesium, while soft water's principal mineral is sodium. Soft water adds appreciable amounts of sodium to the diet and may aggravate hypertension and heart disease. Hard water may oppose these conditions because of its calcium content.

2. Because absorption of calcium varies not only with age, but also with a person's vitamin D status and the calcium content of the diet.

3. Because individuals develop their peak bone mass, which is the highest attainable bone density for an individual, during childhood and adolescence. By the late twenties, the skeleton no longer adds significantly to bone density and the bones begin to lose density.

4. Diuretics are medications which cause increased water excretion and dehydration which leads to potassium loss from inside cells. It is especially dangerous because potassium loss from brain cells makes the victim unaware of the need for water. Potassium-rich foods are necessary to compensate for the losses.

5. Pica is a craving for non-food substances, such as ice, clay or paste, and is associated with iron deficiency.

6. DASH stands for Dietary Approaches to Stop Hypertension and it helps to lower blood pressure. The diet emphasizes increased intakes of fruits and vegetables and adequate intakes of nuts, fish, whole grains, and low-fat dairy products. Red meat, butter, other high-fat foods, sweets and salt and sodium are reduced.

Short Answer Questions:

1. (a) calcium; (b) chloride; (c) magnesium; (d) phosphorus; (e) potassium; (f) sodium; (g) sulfur

2. (a) calcium; (b) magnesium; (c) sodium

3. (a) regulates transport of ions across cell membranes and is important in nerve transmission; (b) helps maintain normal blood pressure; (c) is essential for muscle contraction and maintenance of heartbeat; (d) plays a role in clotting of blood; (e) allows secretion of hormones, digestive enzymes, and neurotransmitters; (f) activates cellular enzymes that regulate many processes

4. (a) overly strict use of low sodium diets; (b) vomiting; (c) diarrhea; (d) heavy sweating

5. (a) learn to control use of the salt shaker; (b) limit intake of processed and fast foods

6. (a) inadequate intake; (b) vomiting; (c) diarrhea; (d) alcoholism; (e) protein malnutrition

7. (a) seafood; (b) iodized salt

8. (a) tannins of tea and coffee; (b) calcium and phosphorus in milk; (c) phytates that accompany fiber in lightly processed whole-grain cereals and legumes

Problem-Solving:

Decrease the amount of salt; substitute ¼ cup oil for the $^{1}/_{3}$ cup lard; use fresh or frozen green peas rather than canned green peas; delete the ¼ cup butter; delete the poultry seasoning; use unsalted and fat-free chicken broth; substitute skim milk for the cream.

Controversy Questions:

1. Trabecular bone is the web-like structure composed of calcium-containing crystals inside a bone's solid outer shell; it provides strength and acts as a calcium storage bank; cortical bone is the dense, ivory-like bone that forms the exterior shell of a bone and the shaft of a long bone; it provides a sturdy outer wall.

2. Poor nutrition involving calcium and vitamin D; estrogen deficiency in women; lack of physical activity; being underweight; use of alcohol and tobacco; and excess protein, sodium, caffeine, and soft drinks and inadequate vitamin K.

3. It is thought that genetic inheritance influences the maximum bone mass possible during growth and the extent of a woman's bone loss during menopause.

4. Weight-bearing exercises such as calisthenics, dancing, jogging, vigorous walking or weight training.

5. Using milk as the primary beverage; playing actively or engaging in sports; limiting television; drinking fluoridated water; and not starting to smoke or drink alcohol.

Study Aids

Major Minerals			
Name	Chief Functions	Deficiency Disease Name	Food Sources
Calcium	• makes up bone and teeth structure • normal muscle contraction • nerve functioning • blood clotting • immune defenses • blood pressure	osteoporosis	• milk • small fish with bones • tofu • greens • legumes
Magnesium	• bone mineralization • building of protein • normal muscle contraction • transmission of nerve impulses • maintenance of teeth		• nuts • legumes • whole grains • dark green vegetables • seafoods • chocolate • cocoa
Sodium	• maintains normal fluid balance • maintains normal acid-base balance		• salt • processed foods • soy sauce

Trace Minerals			
Name	Chief Functions	Deficiency Disease Name	Food Sources
Iron	• part of the protein hemoglobin • part of the protein myoglobin • necessary for utilization of energy in the body	anemia	• red meats • fish • poultry • shellfish • eggs • legumes • dried fruits
Iodine	• component of thyroid hormone thyroxine	goiter cretinism	• seafood • iodized salt • bread
Zinc	• part of insulin and many enzymes • normal fetal development		• protein-containing foods

Sample Test Items

1. c (p. 264)	9. d (p. 299)	17. c (p. 286-287)	25. a (p. 275)
2. d (p. 267)	10. c (p. 279)	18. d (p. 288)	26. e (p. 275)
3. a (p. 268)	11. d (p. 281)	19. a (p. 294)	27. b (p. 299)
4. d (p. 268)	12. c (p. 281)	20. d (p. 284)	28. c (p. 279)
5. c (p. 273)	13. d (p. 281)	21. e (p. 267)	29. e (p. 285)
6. d (p. 275)	14. a (p. 284)	22. c (p. 268)	30. e (p. 290)
7. d (p. 275)	15. e (p. 278)	23. a (p. 277)	
8. a (p. 275)	16. b (p. 285)	24. d (p. 280)	

Chapter 9 - Energy Balance and Healthy Body Weight

Chapter Objectives

After completing this chapter, you should be able to:

1. List and define the three components of the body's energy budget.

2. Identify and explain the factors that affect the basal metabolic rate.

3. Estimate an individual's total energy requirement.

4. Discuss the problems of too much or too little body fat.

5. Describe health risks associated with central obesity and those groups most likely to be affected.

6. Discuss the role of BMI in evaluating obesity.

7. Evaluate the methods used for estimating body fatness.

8. Summarize the theories that attempt to explain the mystery of obesity.

9. Explain what happens during moderate weight loss versus rapid weight loss.

10. Summarize the recommended strategies to promote weight control.

11. Explain strategies that work best for weight gain.

12. Describe how physical activity contributes to a weight loss program.

13. Identify the pros and cons of high-protein, low-carbohydrate diets.

14. Define and describe eating disorders and explain the physical harm that occurs as a result of these behaviors (Controversy 9).

Key Concepts

✓ Deficient body fatness threatens survival during a famine or in wasting diseases.

✓ Most obese people suffer illnesses, and obesity is considered a chronic disease.

✓ Central obesity may be more hazardous to health than other forms of obesity.

✓ Experts estimate health risks from obesity using BMI values, waist circumference, and a disease risk profile. Fit people are healthier than others, regardless of body fatness.

✓ Overfatness presents social and economic handicaps as well as physical ills. Judging people by their body weight is a form of prejudice in our society.

- The "energy in" side of the body's energy budget is measured in calories taken in each day in the form of foods and beverages. The number of calories in foods and beverages can be obtained from published tables or computer diet analysis programs. No easy method exists for determining the "energy out" side of a person's energy balance equation.

- The DRI Committee sets Estimated Energy Requirements for a reference man and woman. People's energy needs vary greatly.

- Two major components of the "energy out" side of the body's energy budget are basal metabolism and voluntary activities. A third minor component of energy expenditure is the thermic effect of food. Many factors influence the basal metabolic rate.

- The DRI Committee established a method for determining an individual's approximate energy requirement.

- The BMI values mathematically correlate heights and weights with risks to health. They are especially useful for evaluating health risks of obesity but fail to measure body composition or fat distribution.

- A clinician can determine the percentage of fat in a person's body by measuring fatfolds, body density, or other parameters. Distribution of fat can be estimated by radiographic techniques, and central adiposity can be assessed by measuring waist circumference.

- No single body composition or weight suits everyone; needs vary by gender, lifestyle, and stage of life.

- Hunger is stimulated by an absence of food in the digestive tract. Appetite can occur with or without hunger.

- Satiation occurs when the digestive organs signal the brain that enough food has been consumed. Satiety is the feeling of fullness that lasts until the next meal. Hunger outweighs satiety in the appetite control system.

- The adipose tissue hormone leptin suppresses the appetite in response to a gain in body fat.

- Some foods may confer greater satiety than others.

- Metabolic theories attempt to explain obesity on the basis of molecular functioning.

- A person's genetic inheritance greatly influences the likelihood of obesity. Genetic researchers are identifying potential genetic links with obesity development.

- Studies of human behavior identify stimuli that lead to overeating. Food pricing, availability, and advertising influence food choices. Physical inactivity is clearly linked with overfatness.

- When energy balance is negative, glycogen returns glucose to the blood. When glycogen runs out, body protein is called upon for glucose. Fat also supplies fuel as fatty acids. If glucose runs out, fat supplies fuel as ketone bodies, but ketosis can be dangerous. Both prolonged fasts and low-carbohydrate diets are ill-advised.

- When energy balance is positive, carbohydrate is converted to glycogen or fat, protein is converted to fat, and food fat is stored as fat. Alcohol delivers calories and encourages fat storage.

- To achieve and maintain a healthy body weight, set realistic goals, keep records, and expect to progress slowly. Watch energy density, make the diet adequate and balanced, limit calories, reduce alcohol, and eat regularly, especially at breakfast.

- Physical activity greatly augments diet in weight loss efforts. Improvements in health and body composition follow an active lifestyle.

✓ Weight gain requires a diet of calorie-dense foods, eaten frequently throughout the day. Physical activity builds lean tissue, and no special supplements can speed the process.

✓ For people whose obesity threatens their health, medical science offers drugs and surgery. The effectiveness of herbal products and other gimmicks has not been demonstrated, and they may prove hazardous.

✓ People who succeed at maintaining lost weight keep to their eating routines, keep exercising, and keep track of calorie and fat intakes and body weight. The more traits related to positive self-image and self-efficacy a person possesses or cultivates, the more likely that person will succeed.

Summing Up

Both deficient and (1)_____ body fat present health risks. For example, underweight increases the risk for any person fighting a (2)_____ disease, while overfatness can precipitate (3)_____ and thus increase the risk of stroke. Being overfat also triples a person's risk of developing (4)_____. The (5)_____ usually correlates with degree of body fatness and disease risks.

The "energy in" side of the body's energy budget is measured in (6)_____ taken in each day in the form of foods and beverages. Calories in foods and beverages can be obtained from published tables or (7)_____ programs. The body spends energy in two major ways: to fuel its (8)_____ metabolism and to fuel its (9)_____ activities.

The DRI Committee sets (10)_____ values by taking into account the ways in which energy is spent and by whom. A clinician can determine the percentage of fat in a person's body by measuring fatfolds, body (11)_____, or other parameters. Distribution of fat can be estimated by radiographic techniques and central adiposity by measuring (12)_____ circumference.

Different theories exist to explain the development of obesity, with one attributing it to internal factors and the other to (13)_____ factors. One metabolic theory suggests that the body chooses a weight it wants to maintain and defends that weight, and is referred to as the (14)_____ theory. A different line of research contends that obesity is determined by behavioral responses to environmental stimuli, which forms the rationale for the importance of (15)_____ cues to overeating. Physical (16)_____ is clearly linked with overfatness.

Researchers are interested in why people eat when and what they eat, and especially why some people overeat, and this has lead to investigations of hunger, (17)_____, and satiety. Hunger is the (18)_____ need to eat, while appetite is the (19)_____ desire to eat. The perception of (20)_____ that lingers in the hours after a meal is called satiety.

(21)_____ is a hormone produced by the adipose tissue that is directly linked to both appetite control and body fatness. Of the three energy-yielding nutrients, (22)_____ is perhaps the most satiating.

Changes in body weight can reflect shifts in body fluid content, in bone minerals, or in (23)_____ tissues such as muscles. The type of (24)_____ gained or lost depends on how the person goes about gaining or losing it. With prolonged fasting or carbohydrate deprivation, the body adapts by converting (25)_____ into compounds that the nervous system can adapt to use and this condition is called (26)_____.

Many different approaches to weight loss are used including fasting and low-(27)_____ diets which are ill-advised. People who successfully lose and maintain weight loss set realistic goals, keep (28)_____, and expect to progress slowly.

Chapter Glossary

Crossword Puzzle:

Across:	Down:
5. measurement of the thickness of a fold of skin on the back of the arm (over the triceps muscle), below the shoulder blade (subscapular), or in other places, using a caliper	1. fat stored within the abdominal cavity in association with the internal abdominal organs
6. intentional activities conducted by voluntary muscles	2. the sum total of all the involuntary activities that are necessary to sustain life, including respiration, circulation, and new tissue synthesis, but excluding digestion and voluntary activities
11. the rate at which the body uses energy to support its basal metabolism	3. an indicator of obesity, calculated by dividing the weight of a person by the square of the person's height
12. a measure of density and volume used to determine body fat content	4. a term popularly used to describe dimpled fat tissue on the thighs and buttocks; not recognized in science
14. overfatness, with adverse health effects	7. compounds of the brain whose actions mimic those of opiate drugs in reducing pain and producing pleasure
15. excess fat in the abdomen and around the trunk	8. fat stored directly under the skin
	9. a person's belief in his or her ability to succeed in an undertaking
	10. repeated rounds of weight loss and subsequent regain, with reduced ability to lose weight with each attempt
	13. a noninvasive method of determining total body fat, fat distribution and bone density

basal metabolic rate	DEXA	subcutaneous fat
basal metabolism	endorphins	underwater weighing
body mass index	fatfold test	visceral fat
cellulite	obesity	voluntary activities
central obesity	self-efficacy	weight cycling

Matching Exercise:

_____ 1. leptin

_____ 2. ghrelin

_____ 3. thermogenesis

_____ 4. wasting

_____ 5. appetite

_____ 6. thermic effect of food

_____ 7. satiation

_____ 8. set-point theory

_____ 9. adipose tissue

_____ 10. ketone bodies

_____ 11. body composition

_____ 12. extreme obesity

_____ 13. lipoprotein lipase

_____ 14. hunger

_____ 15. bioelectrical impedance

_____ 16. energy density

_____ 17. overweight

_____ 18. underweight

a. the body's speeded-up metabolism in response to having eaten a meal

b. acidic compounds derived from fat and certain amino acids

c. a technique to measure body fatness by measuring the body's degree of electrical conductivity

d. an appetite-suppressing hormone produced in the fat cells that conveys information about body fatness to the brain

e. the theory that the body tends to maintain a certain weight by means of its own internal controls

f. an enzyme mounted on the surfaces of fat cells that splits triglycerides in the blood into fatty acids and glycerol to be absorbed into the cells for reassembly and storage

g. the generation and release of body heat associated with the breakdown of body fuels

h. the body's fat tissue, which performs several functions

i. a hormone released by the stomach that signals the hypothalamus to stimulate eating

j. the psychological desire to eat; a learned motivation and a positive sensation that accompanies the sight, smell, or thought of appealing foods

k. the proportions of muscle, bone, fat, and other tissue that make up a person's total body weight

l. a measure of the energy provided by a food relative to its weight

m. the physiological need to eat, experienced as a drive for obtaining food; an unpleasant sensation that demands relief

n. the progressive, relentless loss of the body's tissues that accompanies certain diseases and shortens survival time

o. the perception of fullness that builds throughout a meal, eventually reaching the degree of fullness and satisfaction that halts eating

p. the condition of having a Body Mass Index of 40 or above

q. having a Body Mass Index of less than 18.5

r. having a Body Mass Index of 25.0 – 29.9

Exercises

Answer these chapter study questions:

1. What are risks associated with being underweight?

2. What are two major drawbacks of BMI values?

3. Identify appropriate strategies for weight gain.

4. Describe diet-related changes that most often lead to successful weight change and maintenance.

5. Describe how exercise can increase the BMR.

6. What factors are considered by the DRI Committee in setting Estimated Energy Requirements (EER)?

Complete these short answer questions:

1. The FDA's plan to confront the obesity epidemic includes:

   a.
   b.
   c.
   d.
   e.

2. Two major components of the "energy out" side of the body's energy budget are:

   a.
   b.

3. Body mass index is of limited use with:

   a.
   b.
   c.

4. Techniques for estimating body fatness include:

   a
   b.
   c.
   d.

5. Leptin is a hormone produced by adipose tissue that is directly linked to:

   a.
   b.

6. Gradual weight loss is preferred to rapid weight loss because:

   a.
   b.

7. Changes in body weight may reflect shifts in any of the following materials:

   a.
   b.
   c.
   d.

8. The three kinds of energy nutrients are stored in the body in two forms, including _____ and _____.

9. Benefits of physical activity in a weight-management program include:

   a.   e.
   b.   f.
   c.   g.
   d.

10. In late food deprivation, _____ bodies help feed the nervous system and so help spare tissue protein.

Solve these problems:

1. Calculate the estimated energy requirement for a 110 pound female who is 5' 3" tall and is sedentary. She is 25 years old.

2. Using the U.S. Food Exchange System in Appendix D of your textbook, calculate the number of calories provided by the following meal:

Food Item	Amount	Number of Calories
Boiled egg	1	_____
Whole wheat toast	1 slice	_____
Margarine	2 teaspoons	_____
Grapefruit	½	_____
Nonfat milk	1 cup	_____
Total:		_____

3. Calculate the grams of carbohydrate, protein, and fat which should be supplied in a 1000 calorie diet, using the distribution of 50% carbohydrate, 25% protein, and 25% fat.

Answer these controversy questions:

1. What are the three associated medical problems that form the female athlete triad?

2. What are three characteristics specific to anorexia nervosa?

3. What two issues and behaviors must be addressed in the treatment of anorexia nervosa?

4. What are characteristics of typical binge foods that are consumed during a bulimic binge?

5. What are the two primary goals of treatment for bulimia nervosa?

Study Aids

1. Identify whether the following factors are associated with a higher or lower *Basal Metabolic Rate (BMR)* by placing an X in the appropriate space.

	Factor	*Higher BMR*	*Lower BMR*
a.	Youth	_____	_____
b.	Being tall and thin	_____	_____
c.	Starvation	_____	_____
d.	Heat	_____	_____
e.	Fasting	_____	_____
f.	Stress hormones	_____	_____
g.	Pregnant women	_____	_____
h.	Fever	_____	_____
i.	Thyroxine	_____	_____

2. Fill in the missing terms in this diagram on feasting and fasting.

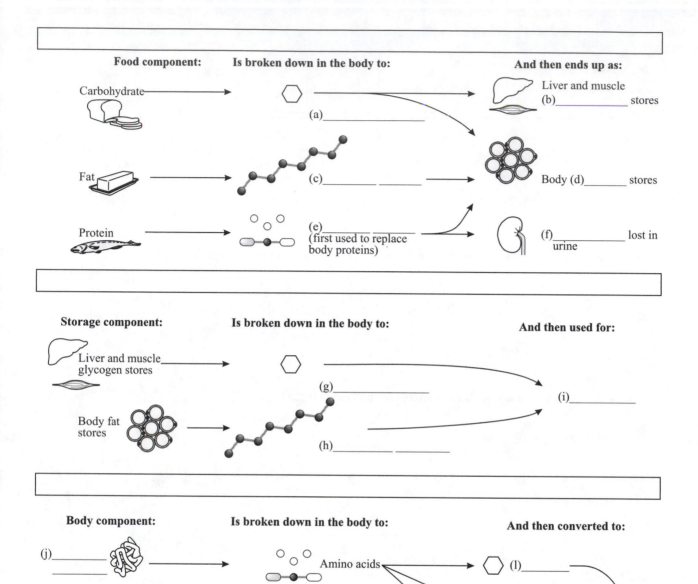

Food component: Is broken down in the body to: And then ends up as:

Carbohydrate ———————→   ⬡ ——————————→   Liver and muscle
                                              (a)_____   (b)_____ stores

Fat ————————→   (c)_____ _____ ——→   Body (d)_____ stores

Protein ————————→   (e)_____   (f)_____ lost in
                                              (first used to replace   urine
                                              body proteins)

Storage component: Is broken down in the body to: And then used for:

Liver and muscle ——————→   ⬡ ————————
glycogen stores   (g)_____   (i)_____

Body fat ——————→
stores   (h)_____ _____

Body component: Is broken down in the body to: And then converted to:

(j)_____ ————————→   Amino acids ⬡ (l)_____
   _____   (m)_____
                                                                                                and some   Energy
                                                                                                ketone bodies
                                                                                                lost in urine
(k)_____ ——————→   Fatty acids ——————→   (n)_____
   _____   _____

144

Sample Test Items

Comprehension Level Items:

1. Health risks associated with excessive body fat include all of the following **except:**

    a. increased risk of developing hypertension.
    b. increased risk for developing ulcers.
    c. increased risk for heart disease.
    d. increased risk for developing diabetes.

2. Which of the following groups is **least** likely to carry more intraabdominal fat?

    a. men
    b. women in their reproductive years
    c. women past menopause
    d. smokers

3. Obesity is officially defined as a body mass index of 30 or higher.

    a. true
    b. false

4. Outside-the-body causes of obesity suggest that:

    a. people eat in response to internal factors.
    b. people eat in response to factors such as hunger.
    c. environmental influences override internal regulatory systems.
    d. the body tends to maintain a certain weight by means of external controls.

5. A man of normal weight may have, on the average, _____ percent of the body weight as fat.

    a. 5-10
    b. 10-15
    c. 12-20
    d. 20-30

6. BMI values are most valuable for evaluating degrees of obesity and are less useful for evaluating nonobese people's body fatness.

    a. true
    b. false

7. Central obesity correlates with an increased incidence of:

    a. diabetes.
    b. kidney disease.
    c. stroke.
    d. a and b
    e. a and c

145

8. The theory that the body tends to maintain a certain weight by means of its own internal controls is called the _____ theory.

   a. enzyme
   b. set-point
   c. thermogenesis
   d. fat cell

9. Which of the following helps to feed the brain during times when too little carbohydrate is available?

   a. lipoprotein lipase
   b. brown fat
   c. ketone bodies
   d. leptin

10. Natural herbs are never harmful to the body.

    a. true
    b. false

11. Which of the following largely determines whether weight is gained as body fat or as lean tissue?

    a. the number of calories consumed
    b. the amount of fat consumed
    c. the person's current body composition
    d. exercise

12. Hunger makes itself known roughly _____ hours after eating.

    a. 2 - 4
    b. 4 - 6
    c. 8 - 12
    d. 12 - 14

13. Ketone bodies are:

    a. acidic compounds.
    b. derived from fat.
    c. normally present in the blood.
    d. a and b
    e. b and c

14. Which of the following directly controls basal metabolism?

    a. epinephrine
    b. insulin
    c. thyroxine
    d. lipase

15. Which of the following statements is **not** true?

    a. Any food can make you fat if you eat enough of it.
    b. Alcohol both delivers calories and encourages fat accumulation.
    c. Fat from food, as opposed to carbohydrate or protein, is especially easy for the body to store as fat tissue.
    d. Protein, once converted to fat, can later be recovered as amino acids.

16. Diet strategies for weight loss should include all of the following **except**:

    a. skipping meals.
    b. using the concept of energy density.
    c. choosing realistic calorie levels.
    d. keeping records.

17. Fatfold measurements provide an accurate estimate of total body fat and a fair assessment of the fat's location.

    a. true
    b. false

18. Which of the following is the first element in implementing a behavior modification program?

    a. Reward yourself personally and immediately.
    b. Strengthen cues to appropriate behaviors.
    c. Eliminate inappropriate eating cues.
    d. Arrange negative consequences for negative behaviors.

19. All of the following are considered to be part of basal metabolism **except:**

    a. beating of the heart.
    b. inhaling and exhaling of air.
    c. maintenance of body temperature.
    d. sitting and walking.

20. Fat is the most satiating of the three energy-yielding nutrients.

    a. true
    b. false

21. Gradual weight loss is preferred to rapid weight loss because:

    a. lean body mass is spared.
    b. fat is lost.
    c. lean body mass is lost.
    d. a and b
    e. b and c

22. Most weight loss can lead to rapid improvements in control over diabetes, blood pressure, and blood lipids.

    a. true
    b. false

23. According to a diet evaluation by scientists from USDA, most people who recently lost substantial weight were those eating diets:

    a. high in carbohydrates.
    b. low in fat.
    c. low in carbohydrates.
    d. a and b
    e. b and c

24. The DRI Committee recommends at least _____ grams of carbohydrates a day.

    a. 120
    b. 125
    c. 130
    d. 135

Application Level Items:

25. Kathy weighs 140 pounds and has a body mass index of 28. Based on this information you would classify her as:

    a. underweight.
    b. normal weight.
    c. overweight.
    d. obese.

26. John has a body mass index of 31; therefore, you would classify him as:

    a. underweight.
    b. normal weight.
    c. overweight.
    d. obese, class I.

27. Susie is underweight and wishes to gain weight. Which of the following should she do?

    a. exercise
    b. eat only three meals a day
    c. eat a high-calorie diet
    d. a and b
    e. a and c

28. Emily has been overweight most of her life and she has been on almost every type of diet possible. Although she loses weight initially on reducing diets, she subsequently returns to her original weight. Which of the following theories could explain Emily's problem with being overweight?

    a. fat cell number theory
    b. set-point theory
    c. adaptive thermogenesis theory
    d. enzyme theory

29. Sarah is bragging about losing 10 pounds in one week on a weight loss program. What would you tell Sarah?

    a. It's great that you have lost that much weight so fast.
    b. You need to slow down your rate of weight loss.
    c. The weight loss probably represents changes in body fluid content.
    d. Your change in weight probably reflects a loss of body fat.

30. Karen is interested in speeding up her basal metabolic rate to promote fat loss. You would advise her to:

    a. eat smaller meals more frequently throughout the day.
    b. participate in endurance and strength-building exercise daily.
    c. fast at least one day each week.
    d. follow a low-carbohydrate diet.

Answers

Summing Up

(1) excessive; (2) wasting; (3) hypertension; (4) diabetes; (5) Body Mass Index; (6) calories; (7) computer diet analysis; (8) basal; (9) voluntary; (10) Estimated Energy Requirement; (11) density; (12) waist; (13) external; (14) set-point; (15) external; (16) inactivity; (17) appetite; (18) physiological; (19) psychological; (20) fullness; (21) leptin; (22) protein; (23) lean; (24) tissue; (25) fat; (26) ketosis; (27) carbohydrate; (28) records

Chapter Glossary

Crossword Puzzle: (1) visceral fat; (2) basal metabolism; (3) body mass index; (4) cellulite; (5) fatfold test; (6) voluntary activities; (7) endorphins; (8) subcutaneous fat; (9) self-efficacy; (10) weight cycling; (11) basal metabolic rate; (12) underwater weighing; (13) DEXA; (14) obesity; (15) central obesity
Matching Exercise: (1) d; (2) i; (3) g; (4) n; (5) j; (6) a; (7) o; (8) e; (9) h; (10) b; (11) k; (12) p; (13) f; (14) m; (15) c; (16) l; (17) r; (18) q

Exercises

Chapter Study Questions:

1. Overly thin people are at a disadvantage in the hospital where they may have to refrain from eating food for days at a time so that they can undergo tests or surgery and their nutrient status can easily deteriorate. Underweight also increases the risk for any person fighting a wasting disease such as cancer.

2. They fail to indicate how much of the weight is fat and where the fat is located.

3. Participate in physical activity, especially strength training; increase calories and learn to eat nutritious, calorie-dense foods; and eat more frequently.

4. Keep records of food intake and exercise habits; set realistic goals; plan a diet with foods you like; pay attention to portions; eat regularly; include breakfast; cut down on or eliminate alcohol; watch energy density; make the diet adequate and balanced with carbohydrates, protein, and fat; limit calories.

5. Exercise helps body composition change toward the lean. Lean tissue is more metabolically active than fat tissue, so the basal energy output picks up the pace as well.

6. A person's gender, age, physical activity and body size and weight are considered in estimating a person's EER.

Short Answer Questions:

1. (a) requiring food labels to display calorie counts prominently and to use meaningful serving sizes; (b) launching a calories count consumer education campaign; (c) encouraging restaurants to provide nutrition information to consumers; (d) encouraging greater accuracy of food label nutrient data; (e) encouraging development of obesity drugs and stepping up research activities

2. (a) basal metabolism; (b) voluntary activities

3. (a) athletes; (b) adults over 65; (c) pregnant and lactating women

4. (a) anthropometry; (b) density; (c) conductivity; (d) radiographic techniques

5. (a) appetite control; (b) body fatness

6. (a) lean body mass is spared; (b) fat is lost

7. (a) fluid content; (b) bone minerals; (c) lean tissue; (d) contents of the bladder or digestive tract

8. glycogen; fat

9. (a) short-term increase in energy expenditure; (b) long-term increase in BMR; (c) improved body composition; (d) appetite control; (e) stress reduction and control of stress eating; (f) physical and psychological well-being; (g) improved self-esteem

10. ketone

Problem-Solving:

1. (a) change pounds to kilograms: 110 pounds divided by 2.2. pounds = 50 kilograms;
    (b) change height in feet and inches to inches, and then to meters: 63 inches ÷ 39.37 = 1.6 meters;
    (c) insert numbers into the equation for women 19 years and older:

$$EER = 662 - 6.91 \times age + PA \times [(9.36 \times wt) + (726 \times ht)] =$$
$$662 - 6.91 \times 25 + 1.0 \times [(9.36 \times 50) + (726 \times 1.6)] =$$
$$662 - 6.91 \times 25 + 1.0 \times [468 + 1162] =$$
$$662 - 6.91 \times 25 + 1.0 \times 1630 =$$
$$662 - 172.75 + 1630 = \quad 2119 \text{ calories per day}$$

2.

Food Item	*Amount*	*Number of Calories*
Boiled egg	1	75
Whole wheat toast	1 slice	80
Margarine	2 teaspoons	90
Grapefruit	½	60
Nonfat milk	1 cup	90
Total:		395

3. (a) 50% carbohydrate × 1000 calories = 500 carbohydrate calories divided by 4 calories per gram = 125 grams carbohydrate;
    (b) 25% protein × 1000 calories = 250 protein calories divided by 4 calories per gram = 63 grams protein;
    (c) 25% fat × 1000 calories = 250 fat calories divided by 9 calories per gram = 28 grams fat

Controversy Questions:

1. Disordered eating, amenorrhea, and osteoporosis.

2. A refusal to maintain a body weight at or above a minimal normal weight for age and height; intense fear of gaining weight or becoming fat; influence of body weight or shape on self-evaluation.

3. Those relating to food and weight and those involving relationships with oneself and others.

4. Typical foods are easy-to-eat, low-fiber, smooth-textured, high-fat, and high-carbohydrate foods.

5. Steady maintenance of weight and prevention of relapse into cyclic gains and losses.

Study Aids

1.
Factor	Higher BMR	Lower BMR
a. Youth	x	
b. Being tall and thin	x	
c. Starvation		x
d. Heat	x	
e. Fasting		x
f. Stress hormones	x	
g. Pregnant women	x	
h. Fever	x	
i. Thyroxine	x	

2. (a) glucose; (b) glycogen; (c) fatty acids; (d) fat; (e) amino acids; (f) nitrogen; (g) glucose; (h) fatty acids; (i) energy; (j) body protein; (k) body fat; (l) glucose; (m) nitrogen; (n) ketone bodies

Sample Test Items

1. b (p. 313)
2. b (p. 313-314)
3. a (p. 314)
4. c (p. 327-328)
5. c (p. 322)
6. a (p. 320)
7. e (p. 313)
8. b (p. 326)
9. c (p. 330)
10. b (p. 346)

11. d (p. 343)
12. b (p. 322)
13. d (p. 330)
14. c (p. 317)
15. d (p. 333)
16. a (p. 336-342)
17. a (p. 320)
18. c (p. 348)
19. d (p. 317)
20. b (p. 325)

21. d (p. 330)
22. a (p. 333)
23. d (p. 334-335)
24. c (p. 332)
25. c (p. 314)
26. d (p. 314)
27. e (p. 343-344)
28. b (p. 326)
29. c (p. 329-332)
30. b (p. 342)

Chapter 10 - Nutrients, Physical Activity, and the Body's Responses

Chapter Objectives

After completing this chapter, you should be able to:

1. Explain the benefits of and guidelines for regular physical activity for the body.

2. Describe the components of fitness and how muscles respond to physical activity.

3. Summarize how the body adjusts its fuel mix to respond to physical activity of varying intensity levels and duration.

4. Explain the concept of carbohydrate loading.

5. Discuss the roles vitamins and minerals play in physical performance and indicate whether supplements are necessary to support the needs of active people.

6. Describe the importance of fluids and temperature regulation in physical activity, and the best way to stay hydrated before and during exercise.

7. Describe the characteristics of the most appropriate diet for athletes.

8. List the risks of taking "ergogenic" aids and steroids to enhance physical performance (Controversy 10).

Key Concepts

✓ Physical activity and fitness benefit people's physical and psychological well-being and improve their resistance to disease. Physical activity to improve physical fitness offers additional personal benefits.

✓ The components of fitness are flexibility, muscle strength, muscle endurance, and cardiorespiratory endurance. To build fitness, a person must engage in physical activity. Muscles adapt to activities they are called upon to perform repeatedly.

✓ Weight training offers health and fitness benefits to adults. Weight training reduces the risk of cardiovascular disease, improves older adults' physical mobility, and helps maximize and maintain bone mass.

✓ Cardiorespiratory endurance training enhances the ability of the heart and lungs to deliver oxygen to the muscles. With cardiorespiratory endurance training, the heart becomes stronger, breathing becomes more efficient, and the health of the entire body improves.

✓ Glucose is supplied by dietary carbohydrate or made by the liver. It is stored in both liver and muscle tissue as glycogen. Total glycogen stores affect an athlete's endurance.

✓ The more intense an activity, the more glucose it demands. During anaerobic metabolism, the body spends glucose rapidly and accumulates lactic acid.

✓ Physical activity of long duration places demands on the body's glycogen stores. Carbohydrate ingested before and during long-duration activity may help to forestall hypoglycemia and fatigue. Carbohydrate loading is a regimen of physical activity and diet that enables an athlete's muscles to store larger-than-normal amounts of glycogen to extend endurance. After strenuous training, eating foods with a high glycemic index may help restore glycogen most rapidly.

✓ Highly trained muscles use less glucose and more fat than do untrained muscles to perform the same work, so their glycogen lasts longer.

✓ Athletes who eat high-fat diets may burn more fat during endurance activity, but the risks to health outweigh any possible performance benefits. The intensity and duration of activity, as well as the degree of training, affect fat use.

✓ Physical activity stimulates muscle cells to break down and synthesize protein, resulting in muscle adaptation to activity. Athletes use protein both for building muscle tissue and for energy. Diet, intensity and duration of activity, and training affect protein use during activity.

✓ Although athletes need more protein than sedentary people, a balanced, high-carbohydrate diet provides sufficient protein to cover an athlete's needs.

✓ Vitamins are essential for releasing the energy trapped in energy-yielding nutrients and for other functions that support physical activity. Active people can meet their vitamin needs if they eat enough nutrient-dense foods to meet their energy needs. Vitamin E may protect against oxidative stress, but there is little evidence that it can improve performance. Iron-deficiency anemia impairs physical performance because iron is the blood's oxygen handler. Sports anemia is a harmless temporary adaptation to physical activity.

✓ Evaporation of sweat cools the body. Heat stroke can be a threat to physically active people in hot, humid weather. Hypothermia threatens those who exercise in the cold.

✓ Physically active people lose fluids and must replace them to avoid dehydration. Thirst indicates that water loss has already occurred.

✓ Water is the best drink for most physically active people, but endurance athletes need drinks that supply glucose as well as fluids.

✓ The body adapts to compensate for sweat losses of electrolytes. Athletes are advised to use foods, not supplements, to make up for these losses.

✓ During events lasting longer than 3 hours, athletes need to pay special attention to replacing sodium losses to prevent hyponatremia.

✓ Caffeine-containing drinks within limits may not impair performance, but water and fruit juice are preferred. Alcohol use can impair performance in many ways and is not recommended.

Summing Up

Physical activity benefits people's physical and psychological well-being and improves their resistance to

(1)_____. Fitness is composed of four components including flexibility, muscle

(2)_____, (3)_____ endurance and cardiorespiratory endurance. Muscle cells and

tissues respond to an (4)_____ of physical activity by gaining strength and size, a response called

(5)_____. The kind of exercise that brings about cardiorespiratory endurance is

(6)_____ activity. Weight training builds (7)_____ and develops

(8)_____ strength and endurance.

The stored glucose of muscle (9)_____ is a major fuel for physical activity. How long an

exercising person's glycogen will last depends not only on diet but also partly on the (10)_____ of

the activity. Glycogen depletion usually occurs after about (11)_____ hours of vigorous exercise. Glucose use

during physical activity depends not only on the intensity, but also on the (12)_____ of the activity.

A person who continues to exercise moderately for longer than 20 minutes begins to use less glucose and more

(13)_____ for fuel. One diet strategy useful for endurance athletes is to eat a high

(14)_____ diet on a day-to-day basis.

Many vitamins and minerals assist in releasing (15)_____ from foods and transporting

(16)_____. Scientists have concluded, however, that nutrient (17)_____ do not enhance the

performance of well-nourished, active people. Vitamin (18)_____ is a potent antioxidant that defends

cell membranes against oxidative damage. Physically active young women, especially those who engage in

endurance activities, are prone to (19)_____ deficiency.

Endurance athletes can lose (20)_____ or more quarts of fluid in every hour of activity and must

hydrate before, and rehydrate (21)_____ and after activity to replace it all. Plain, cool

(22)_____ is the best fluid for most active bodies, although endurance athletes also need

(23)_____ to supplement their limited glycogen stores.

Athletes need a diet composed mostly of (24)_____-dense foods, the kind that supply a

maximum of vitamins and minerals for the energy they provide. A diet that is (25)_____ in

carbohydrate, moderate in (26)_____, and adequate in (27)_____ ensures full

glycogen stores. The (28)_____ meal should be light, easy to (29)_____ and should

contain fluids. In addition, it should be finished at least (30)_____ hours before competition.

Chapter Glossary

Crossword Puzzle:

Word Bank:	
branched-chain amino acids	hypothermia
carbohydrate loading	lactic acid
exercises	muscle strength
glucose polymers	physical activity
heat stroke	pregame meal
hyponatremia	

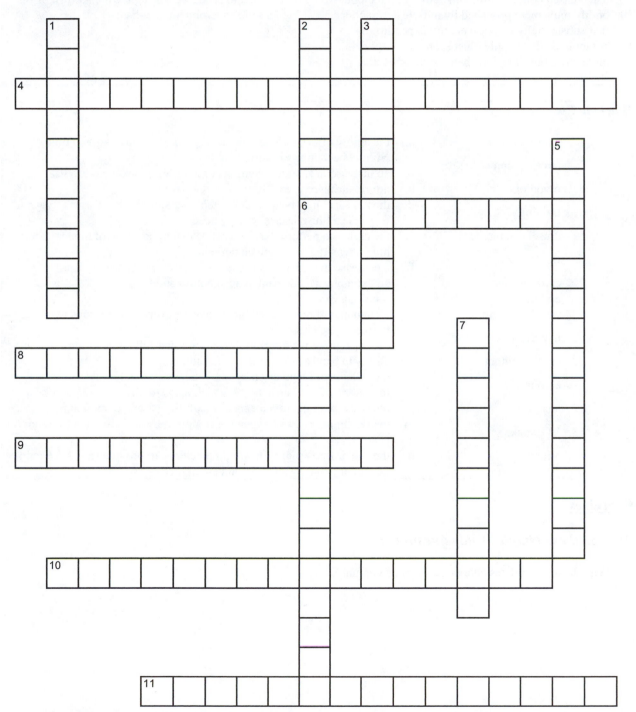

Crossword Puzzle Across:	Crossword Puzzle Down:
4. a regimen of moderate exercise, followed by eating a high-carbohydrate diet, that enables muscles to temporarily store glycogen beyond their normal capacity	1. an acute and life threatening reaction to heat buildup in the body
6. planned, structured, and repetitive bodily movement that promotes or maintains physical fitness	2. the amino acids leucine, isoleucine, and valine, which are present in large amounts in skeletal muscle tissue
8. a below-normal body temperature	3. a meal eaten three to four hours before athletic competition
9. a decreased concentration of sodium in the blood	5. the ability of muscles to work against resistance
10. bodily movement produced by muscle contractions that substantially increase energy expenditure	7. a compound produced during the breakdown of glucose in anaerobic metabolism
11. compounds that supply glucose, not as single molecules, but linked in chains somewhat like starch	

Matching Exercise:

_____	1. overload	a. the ability of a muscle to contract repeatedly within a given time without becoming exhausted
_____	2. stroke volume	b. an increase in size (for example, of a muscle) in response to use
_____	3. myoglobin	c. not requiring oxygen
_____	4. training	d. the muscles' iron-containing protein that stores and releases oxygen in response to the muscles' energy needs
_____	5. muscle endurance	e. the ability to perform large muscle dynamic exercise of moderate to high intensity for prolonged periods
_____	6. flexibility	f. a decrease in size (for example, of a muscle) because of disuse
_____	7. cardiorespiratory endurance	g. an extra physical demand placed on the body
_____	8. weight training	h. requiring oxygen
_____	9. hypertrophy	i. the amount of oxygenated blood ejected from the heart toward body tissues at each beat
_____	10. cardiac output	j. the capacity of the joints to move through a full range of motion; the ability to bend and recover without injury
_____	11. aerobic	k. the volume of blood discharged by the heart each minute
_____	12. anaerobic	l. regular practice of an activity, which leads to physical adaptations of the body, with improvements in flexibility, strength, or endurance
_____	13. VO$_2$ max	m. the maximum rate of oxygen consumption by an individual (measured at sea level)
_____	14. atrophy	n. the use of free weights or weight machines to provide resistance for developing muscle strength and endurance

Exercises

Answer these chapter study questions:

1. Why do some athletes take megadoses of vitamin E?

2. Discuss possible causes of iron deficiency in young female athletes.

3. What is the purpose of carbohydrate loading? Describe the steps involved.

4. What are two routes of significant water loss from the body during exercise?

5. What is the best fluid for most exercising bodies? Why?

6. How would you respond to an athlete who believes that taking vitamin or mineral supplements just before competition will enhance performance?

Complete these short answer questions:

1. The four components of fitness are:

   a.
   b.
   c.
   d.

2. Benefits of weight training include:

   a.
   b.
   c.

3. These vitamins are important for athletes because they build the red blood cells that carry oxygen to working muscles:

   a.
   b.

4. Cardiorespiratory endurance is characterized by:

   a.
   b.
   c.
   d.
   e.
   f.

5. Four strategies which endurance athletes use to try to maintain their blood glucose concentrations as long as they can include:

   a.
   b.
   c.
   d.

6. Factors that affect fat use during physical activity include:

   a.
   b.
   c.

7. During physical activity, the body loses these electrolytes in sweat:

   a.
   b.
   c.

8. Factors that affect how much protein is used during physical activity include:

   a.
   b.
   c.

9. Precautions to prevent heat stroke include:

   a.
   b.
   c.

10. The pregame meal should be eaten _____ hours before athletic competition.

Solve these problems:

1. An athlete weighs 150 pounds prior to a workout and 145 pounds after a workout. How much fluid should be consumed in order to rehydrate the body?

2. Review the example pregame meal below, determine all of the foods and/or beverages which would not be recommended, and identify an appropriate substitute for each non-recommended item.

	**Substitute**
orange juice	_____
3 ounce hamburger patty	_____
bun	_____
potato chips	_____
raw carrot sticks	_____
frozen yogurt	_____

Answer these controversy questions:

1. Explain why large doses of amino acid supplements can be dangerous.

2. What are some of the possible adverse effects of caffeine consumption among athletes?

3. What is the most appropriate way for an athlete to use a specialty drink or bar?

4. Which group of athletes is most likely to use steroids? Why?

5. Why is carnitine supplementation not necessary for an athlete?

Study Aids

For questions 1-5, match the activities, listed on the left, with the type of fitness they develop, listed on the right.

_____1.	weight lifting	a. flexibility
_____2.	repetitive exercises like push-ups	b. muscle strength and endurance
_____3.	stretches	c. cardiorespiratory endurance
_____4.	swimming	
_____5.	fast bicycling	
_____6.	yoga	

For questions 7-11, match the vitamins and minerals, listed on the right, with their exercise-related functions, listed on the left.

_____	7.	helps build the red blood cells that carry oxygen to working muscles		a.	iron
_____	8.	helps make muscles contract		b.	folate
_____	9.	collagen formation for joint and other tissue integrity		c.	calcium
_____	10.	transport of oxygen in blood and in muscle tissue		d.	vitamin C
_____	11.	defends cell membranes against oxidation damage		e.	vitamin E

12. Use Table 10-1 on page 362 of your textbook to study the latest guidelines for physical fitness, and study Table 10-4 on page 377 to identify the hydration schedule for physical activity.

Sample Test Items

Comprehension Level Items:

1. The ability to bend and recover without injury is called:

   a. fitness.
   b. flexibility.
   c. strength.
   d. endurance.

2. People who regularly engage in just moderate physical activity live longer on average than those who are physically inactive.

   a. true
   b. false

3. The average resting pulse rate for adults is around _____ beats per minute.

   a. 20
   b. 50
   c. 70
   d. 90

4. Which of the following produce cardiorespiratory endurance?

   a. swimming
   b. slow walking
   c. basketball
   d. a and b
   e. a and c

5. Sports anemia:

   a. is the same as iron-deficiency anemia.
   b. reflects a normal adaptation to endurance training.
   c. hinders the ability to perform work.
   d. requires iron supplementation.

6. A water loss of _____ percent of body weight can reduce a person's capacity to do muscular work.

   a. 1-2
   b. 3-4
   c. 5-6
   d. 7-8

7. Plain, cool water is the optimal beverage for replacing fluids in those who exercise because it:

   a. rapidly leaves the digestive tract.
   b. takes a long time to enter the tissues.
   c. cools the body from the inside out.
   d. a and b
   e. a and c

8. Which of the following beverages is recommended for an athlete?

   a. iced tea
   b. beer
   c. coffee
   d. fruit juice

9. To postpone exhaustion, endurance athletes should:

   a. eat a high carbohydrate diet on a day-to-day basis.
   b. take in some glucose during exercise.
   c. avoid carbohydrate-rich foods following exercise.
   d. a and b
   e. b and c

10. Which of the following is an effect of regular physical activity?

    a. decreased lean body tissue
    b. less bone density
    c. reduced HDL cholesterol
    d. reduced risk of cardiovascular disease

11 Cardiorespiratory endurance is characterized by:

    a. increased heart strength and stroke volume.
    b. improved circulation.
    c. increased blood pressure.
    d. a and b
    e. b and c

12. The American College of Sports Medicine's guidelines recommend that people spend an accumulated minimum of 30 minutes in some sort of physical activity on most days of the week.

    a. true
    b. false

13. Glycogen depletion usually occurs after about _____ hour(s) of vigorous exercise.

   a. ½
   b. 1
   c. 1 ½
   d. 2

14. During intense activity, anaerobic breakdown of glucose produces:

   a. lactic acid.
   b. ketone bodies.
   c. amino acids.
   d. glycogen.

15. During events lasting longer than 3 hours, athletes need to pay special attention to replacing sodium losses to prevent hyponatremia.

   a. true
   b. false

16. After about _____ minutes of activity, the blood fatty acid concentration rises and surpasses the normal resting concentration.

   a. 10
   b. 20
   c. 30
   d. 45

17. All of the following factors affect fat use in exercise **except:**

   a. amount of fat in the diet.
   b. degree of training to perform the exercise.
   c. intensity and duration of the exercise.
   d. area of the body involved in the exercise.

18. Which of the following directs protein synthesis in building muscle?

   a. protein
   b. nutrients
   c. physical activity
   d. amino acids

19. All of the following are lost in sweat during physical activity **except:**

   a. sodium.
   b. calcium.
   c. potassium.
   d. chloride.

20. Physically fit people have a lower risk of cardiovascular disease due to:

   a. lower blood pressure.
   b. lower HDL cholesterol.
   c. lower total blood cholesterol.
   d. a and b
   e. a and c

21. The Committee on Dietary Reference Intakes recommends at least 60 minutes of moderately intense activity each day to maintain a healthy body weight.

   a. true
   b. false

22. To emphasize muscle strength in weight training, you would combine less resistance (lighter weight) with more repetitions.

   a. true
   b. false

23. To improve your flexibility you would engage in:

   a. weight lifting.
   b. jumping rope.
   c. yoga.
   d. skating.

24. The Committee on DRI recommends greater than normal protein intake for athletes.

   a. true
   b. false

Application Level Items:

25. How would you respond to an athlete who states that branched-chain amino acid supplements should be taken because the branched-chain amino acids can directly provide energy to muscle tissue?

   a. Ordinary foods cannot provide enough branched-chain amino acids.
   b. The body prefers isolated amino acids to the combinations found in food.
   c. When amino acids are needed, the muscles have plenty on hand.
   d. a and b
   e. b and c

26. A male athlete weighs 175 pounds before a race and 168 pounds after a race. How much fluid should he consume?

   a. 8 cups
   b. 10 cups
   c. 12 cups
   d. 14 cups

27. Approximately how many grams of fat would an endurance athlete need to consume to ensure full glycogen and other nutrient stores and to meet energy needs if she consumes 2500 calories?

    a. 40-55 grams
    b. 56-83 grams
    c. 75-95 grams
    d. 100-130 grams

28. Which of the following foods would be most appropriate as part of a pregame meal?

    a. baked potato with chili and sour cream
    b. marinated vegetable salad with whole wheat crackers
    c. salami on rye bread with mayonnaise
    d. pasta with steamed vegetables served with French bread

29. Why are people who exercise in humid, hot weather more susceptible to heat stroke?

    a. because they cannot judge how hot they are getting
    b. because they do not sweat under such conditions
    c. because sweat does not evaporate well under such conditions
    d. a and b
    e. b and c

30. Fred wishes to increase his basal metabolic rate. Which of the following should Fred do?

    a. engage in intense, prolonged activity
    b. drink iced tea
    c. eat a high-carbohydrate diet
    d. drink plenty of water

Answers

Summing Up

(1) disease; (2) strength; (3) muscle; (4) overload; (5) hypertrophy; (6) aerobic; (7) lean body mass; (8) muscle; (9) glycogen; (10) intensity; (11) two; (12) duration; (13) fat; (14) carbohydrate; (15) energy; (16) oxygen; (17) supplements; (18) E; (19) iron; (20) two; (21) during; (22) water; (23) carbohydrate; (24) nutrient, (25) high; (26) fat; (27) protein; (28) pregame; (29) digest; (30) 3-4

Chapter Glossary

Crossword Puzzle: (1) heat stroke; (2) branched-chain amino acids; (3) pregame meal; (4) carbohydrate loading; (5) muscle strength; (6) exercises; (7) lactic acid; (8) hypothermia; (9) hyponatremia; (10) physical activity; (11) glucose polymers
Matching Exercise: (1) g; (2) i; (3) d; (4) l; (5) a; (6) j; (7) e; (8) n; (9) b; (10) k; (11) h; (12) c; (13) m; (14) f

Exercises

Chapter Study Questions:

1. During prolonged, high-intensity physical activity, the muscles' consumption of oxygen increases tenfold or more, enhancing the production of damaging free radicals in the body. Vitamin E is a potent fat-soluble antioxidant that vigorously defends the cell membranes against oxidative damage.

2. Low iron intakes, high iron losses through menstruation, and iron losses through other routes may contribute to iron-deficiency. Iron losses through other routes include the loss of iron through sweat and small blood losses through the digestive tract, and there is also increased iron demand by muscles.

3. It is a technique used to trick the muscles into storing extra glycogen before a competition. During the week prior to competition, the athlete tapers training and then eats a diet that is high in carbohydrates during the three days before competition.

4. Water is lost from the body via sweat, which increases during exercise in order to rid the body of its excess heat by evaporation. In addition, breathing costs water, exhaled as vapor.

5. Plain, cool water because it rapidly leaves the digestive tract to enter the tissues, where it is needed, and it cools the body from the inside out.

6. Vitamins or minerals taken right before an event do not improve performance. Most function as small parts of larger working units and after entering the blood, they have to wait for the cells to combine them with their appropriate other parts so that they can do their work. This may take hours or even longer.

Short Answer Questions:

1. (a) flexibility; (b) muscle strength; (c) muscle endurance; (d) cardiorespiratory endurance

2. (a) reduces risk of cardiovascular disease; (b) improves older adults' physical mobility; (c) helps maximize and maintain bone mass

3. (a) folate; (b) vitamin B_{12}

4. (a) increased cardiac output and oxygen delivery; (b) increased heart strength and stroke volume; (c) slowed resting pulse; (d) increased breathing efficiency; (e) improved circulation; (f) reduced blood pressure

5. (a) eating a high-carbohydrate diet on a day-to-day basis; (b) taking in some glucose during the activity; (c) training the muscles to store as much glycogen as possible; (d) eating carbohydrate-rich foods following activity to boost storage of glycogen

6. (a) fat intake; (b) intensity and duration of the activity; (c) degree of training

7. (a) sodium; (b) potassium; (c) chloride

8. (a) carbohydrate intake; (b) intensity and duration of the activity; (c) degree of training

9. (a) drinking enough fluid before and during the activity; (b) resting in the shade when tired; (c) wearing light-weight clothing that encourages evaporation

10. 3-4

Problem-Solving:

1. One pound of body weight equals roughly two cups of fluid; since 5 pounds of weight were lost, approximately 10 cups of fluid should be consumed.

2.   **Substitute**

       orange juice   _____
       3 ounce hamburger patty   lean chicken breast
       bun   _____
       potato chips   baked potato
       raw carrot sticks   steamed green beans
       frozen yogurt   _____

Controversy Questions:

1. Amino acids compete for carriers, and an overdose of one can limit the availability of some other needed amino acids, setting up both a possible toxicity of the supplemented one and a deficiency of the others. Supplements can also lead to digestive disturbances and excess water accumulation in the digestive tract.

2. Adverse effects include stomach upset, nervousness, irritability, headaches, dehydration, and diarrhea; it can also raise blood pressure above normal, making the heart work harder to pump blood to the working muscles.

3. The most appropriate way for an athlete to use specialty drinks and bars is as a pregame meal or between-meal snack.

4. Professional athletes, for which monetary rewards for excellence are high, are most likely to use steroids.

5. Because carnitine is a nonessential nutrient that the body makes for itself, plus it is contained in meat and milk products.

Study Aids

(1) b; (2) b; (3) a; (4) c; (5) c; (6) a; (7) b; (8) c; (9) d; (10) a; (11) e

Sample Test Items

1. b (p. 362)	11. d (p. 364)	21. a (p. 361)
2. a (p. 360)	12. a (p. 361)	22. b (p. 363)
3. c (p. 364)	13. d (p. 368)	23. c (p. 362)
4. e (p. 364)	14. a (p. 367)	24. b (p. 373)
5. b (p. 375)	15. a (p. 379)	25. c (p. 386)
6. a (p. 376)	16. b (p. 371)	26. d (p. 377)
7. e (p. 377)	17. d (p. 370-371)	27. b (p. 380)
8. d (p. 379)	18. c (p. 371-372)	28. d (p. 382-383)
9. d (p. 368)	19. b (p. 377)	29. c (p. 376)
10. d (p. 361)	20. e (p. 361)	30. a (p. 371)

Chapter 11 - Diet and Health

Chapter Objectives

After completing this chapter, you should be able to:

1. Describe the role of nutrition in maintaining a healthy immune system.

2. Define atherosclerosis and identify the risk factors for cardiovascular disease.

3. Describe how hypertension develops and identify the risk factors associated with this disease.

4. Discuss the strategies that can be used to reduce the risks of cardiovascular disease and hypertension.

5. Describe the process by which a cancer develops and explain what is known about the effects of food constituents on cancer development.

6. Explain the primary causes of the obesity epidemic and how it can be reversed (Controversy 11).

Key Concepts

✓ Adequate nutrition is a key component in maintaining a healthy immune system to defend against infectious diseases. Medical nutrition therapy can improve the course of wasting diseases. Both deficient and excessive nutrients can harm the immune system.

✓ The same diet and lifestyle risk factors may contribute to several degenerative diseases. A person's family history and laboratory test results can reveal strategies for disease prevention.

✓ Plaques of atherosclerosis induce hypertension and trigger abnormal blood clotting, leading to heart attacks or strokes. Abnormal vessel spasms can also cause heart attacks and strokes.

✓ Major risk factors for CVD put forth by the National Cholesterol Education Program include age, gender, family history, high LDL cholesterol and low HDL cholesterol, hypertension, diabetes, obesity, physical inactivity, smoking, and an atherogenic diet. Other potential risk factors are under investigation.

✓ High blood cholesterol indicates a risk of heart disease, and diet can contribute to lowering blood cholesterol and thus the risk of heart disease. Dietary measures to lower LDL cholesterol include reducing intakes of saturated fat, *trans* fat, and cholesterol, along with obtaining the fiber, nutrients, and phytochemicals of fruits, vegetables, legumes, and whole grains. Even for people on prescription medication, diet and exercise are key.

✓ Hypertension is silent, progressively worsens atherosclerosis, and makes heart attacks and strokes likely. All adults should know their blood pressure.

✓ Atherosclerosis, obesity, insulin resistance, age, family background, and race contribute to hypertension risks. Prevention and treatment both deserve high-priority efforts.

✓ For most people, a healthy body weight, regular physical activity, consuming little salt and sodium, moderation for those who use alcohol, and a diet high in fruits, vegetables, fish and low-fat dairy products and low in fat work together to keep blood pressure normal.

- ✓ Cancer arises from genetic damage and develops in steps including initiation and promotion, which are thought to be influenced by diet. The body is equipped to handle tiny doses of carcinogens that occur naturally in foods.

- ✓ Diets high in certain fats and red meats are associated with cancer development. Foods containing ample fiber, folate, calcium, many other vitamins and minerals, and phytochemicals, along with an ample intake of fluid, are thought to be protective.

Summing Up

The diseases that afflict people around the world are of two main kinds, including infectious disease and

(1)_____ disease. Nutrition can help strengthen the body's defenses against

(2)_____ diseases by supporting a healthy immune system. With respect to prevention of

(3)_____ diseases, life choices people make can help to at least postpone them and sometimes to

avoid them altogether.

The body's (4)_____ system guards continuously against disease-causing agents. Impaired

immunity opens the way for diseases, diseases impair (5)_____ assimilation, and nutrition status suffers

further. (6)_____ is especially destructive to various immune system organs and

tissues.

People with AIDS and cancer frequently experience malnutrition and (7)_____ away of the body's

tissues. Nutrients cannot cure or reverse the progression of AIDS, but an adequate diet may help to improve

responses to (8)_____ therapy, reduce duration of (9)_____ stays, and promote

greater (10)_____, with an improved quality of life overall.

In contrast to the infectious diseases, the (11)_____ diseases of adulthood tend to have clusters

of suspected contributors known as (12)_____. Among them are environmental,

(13)_____, social, and genetic factors that tend to occur in clusters and interact with each other. A

person who eats a diet too high in saturated fat, salt and calories increases the probabilities of becoming

(14)_____ and of contracting cancer, hypertension, diabetes, atherosclerosis, diverticulosis or other

diseases. Many experts believe that diet accounts for about a (15)_____ of all cases of coronary

heart disease and general trends support the link between diet and (16)_____.

To decide whether certain (17)_____ recommendations are especially important to you, you

should consider your family's (18)_____ history to see which diseases are common to your forbearers. For example, a person whose close relatives suffered with diabetes and heart disease is advised to avoid

becoming (19)_____ and not to smoke.

In addition to diet, age, gender, cigarette smoking, certain diseases and (20)_____

predict CVD development. Some of the major diet-related risk factors for CVD are (21)_____,

hypertension, and high blood (22)_____ and low blood HDL. High (23)_____ and

low HDL correlate directly with risk of heart disease. An atherogenic diet, which is high in saturated fat and trans

fats, and (24)_____, increases LDL cholesterol.

Primary among the risk factors that precipitate or aggravate hypertension are (25)_____, obesity,

and (26)_____. Epidemiological studies have identified several other risk factors that predict

hypertension including (27)_____ and heredity. People with kidney problems or diabetes, those with a

family history of hypertension, African Americans, and persons who are older are especially likely to be

(28)_____ sensitive. One diet which consistently improves blood pressure in many people who follow

it is called the (29)_____ diet.

Some (30)_____ percent of total cancers are influenced by diet. Studies of populations suggest

that low rates of many cancers correlate with intakes of fiber-rich (31)_____, vegetables and whole

grains, particularly (32)_____. Foods containing ample fiber, folate, calcium, many other vitamins

and minerals, and (33)_____, along with an ample intake of fluid, are thought to be protective.

Chapter Glossary

Crossword Puzzle:

Across:	*Down:*
1. the second figure in a blood pressure reading, which reflects the arterial pressure when the heart is between beats	1. chronic, irreversible diseases characterized by degeneration of body organs due in part to such personal lifestyle elements as poor food choices, smoking, alcohol use, and lack of physical activity
4. an embolus that causes sudden closure of a blood vessel	2. the origination or beginning of cancer
7. a thrombus that has grown enough to close off a blood vessel	3. the first figure in a blood pressure reading, which reflects arterial pressure caused by the contraction of the heart's left ventricle
8. diseases that are caused by bacteria, viruses, parasites, and other microbes and can be transmitted from one person to another through air, water, or food; by contact; or through vector organisms such as mosquitoes or fleas	5. the intentional spreading of disease-causing organisms or agricultural pests as a political weapon to produce fear and intimidate others to gain control
9. the large, primary artery that conducts blood from the heart to the body's smaller arteries	6. acquired immune deficiency syndrome; caused by infection with human immunodeficiency virus (HIV), which is transmitted primarily by sexual contact, contact with infected blood, needles shared among drug users, or fluids transferred from an infected mother to her fetus or infant

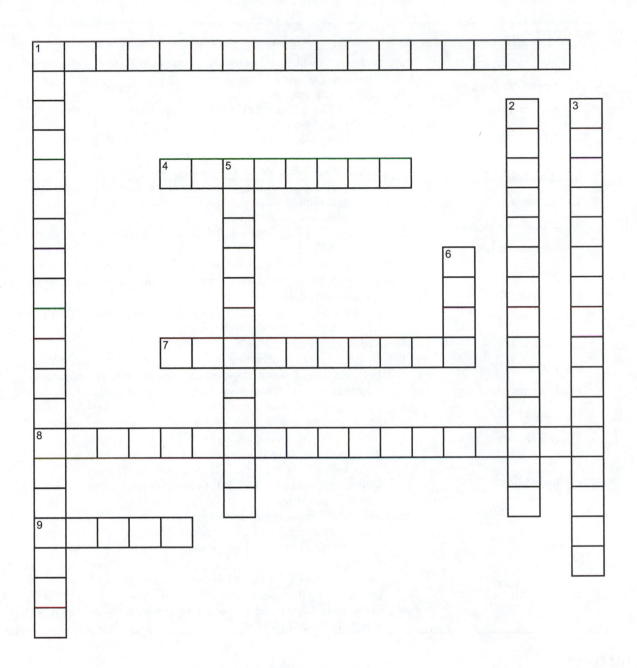

Word Bank:

AIDS
aorta
bioterrorism
carcinogenesis
degenerative diseases

diastolic pressure
embolism
infectious diseases
systolic pressure
thrombosis

Matching Exercise:

_____	1.	anticarcinogens	a.	a disease in which cells multiply out of control and disrupt normal functioning of one or more organs

_____ 1. anticarcinogens

_____ 2. metabolic syndrome

_____ 3. hypertension

_____ 4. carcinogen

_____ 5. plaques

_____ 6. thrombus

_____ 7. metastasis

_____ 8. cruciferous vegetables

_____ 9. atherosclerosis

_____ 10. cancer

_____ 11. calorie effect

_____ 12. heart attack

_____ 13. embolus

_____ 14. aneurysm

_____ 15. initiation

_____ 16. platelets

_____ 17. risk factors

_____ 18. stroke

_____ 19. promoters

a.   a disease in which cells multiply out of control and disrupt normal functioning of one or more organs

b.   high blood pressure

c.   tiny cell-like fragments in the blood, important in blood clot formation

d.   a stationary blood clot

e.   factors that do not initiate cancer but speed up its development once initiation has taken place

f.   a group of vegetables named for their cross-shaped blossoms and associated with low cancer rates

g.   a thrombus that breaks loose

h.   compounds in foods that act in several ways to oppose the formation of cancer

i.   the event in which the vessels that feed the heart muscle become closed off by an embolism, thrombus, or other cause with resulting sudden tissue death

j.   the sudden shutting off of the blood flow to the brain by a thrombus, embolism, or the bursting of a vessel

k.   a cancer-causing substance

l.   an event, probably occurring in the cell's genetic material, caused by radiation or by a chemical carcinogen that can give rise to cancer

m.   a combination of characteristic factors—central obesity, high fasting blood glucose or insulin resistance, hypertension, low blood HDL cholesterol and elevated triglycerides—that greatly increase a person's risk of developing CVD

n.   the drop in cancer incidence seen whenever intake of food energy is restricted

o.   factors known to be related to (or correlate with) diseases but not proved to be causal

p.   mounds of lipid material, mixed with smooth muscle cells and calcium, that develop in the artery walls in atherosclerosis

q.   the ballooning out of an artery wall at a point that is weakened by deterioration

r.   the most common form of cardiovascular disease, characterized by plaques along the inner walls of the arteries

s.   movement of cancer cells from one body part to another, usually by way of the body fluids

Exercises

Answer these chapter study questions:

1.   Why do malnourished children have repeated digestive tract infections?

2. Why must careful attention be paid to food safety when serving food to people with AIDS?

3. How would you go about deciding whether certain diet recommendations and lifestyle changes are especially important to you?

4. What is the difference between a heart attack and a stroke?

5. Why is it erroneous to think that vitamin supplements can provide all of the benefits of vegetables and fruits as related to cancer prevention?

6. Describe the primary features of metabolic syndrome.

Complete these short answer questions:

1. These groups of people are especially likely to be caught in the downward spiral of malnutrition and weakened immunity:

   a.
   b
   c.
   d.

2. The major categories of risk factors for degenerative diseases of adulthood include:

   a.
   b
   c.
   d.

3. The major risk factors for CVD that cannot be modified include:

   a.
   b
   c.

4. The primary risk factors that precipitate or aggravate hypertension are:

   a.
   b
   c.

5. The steps in cancer development are thought to be:

   a.
   b
   c.
   d.
   e.
   f.

6. Alcohol intake is associated with cancers of the:

   a.
   b.
   c.

7. The primary features of metabolic syndrome include:

   a.
   b.
   c.
   d.

8. Examples of cruciferous vegetables include:

    a.
    b
    c.
    d.
    e.
    f.

9. These types of meats should be consumed only occasionally to minimize risks from carcinogens formed during cooking:

    a.
    b
    c.
    d.
    e.

10. For most people, these nutrition measures work to keep blood pressure normal:

    a.
    b
    c.
    d.
    e.

Solve these problems:

1. A serving of a food contains 70 calories and 2 grams of fat. Calculate the percentage of calories from fat in a serving of the food. Would this food be appropriate on a diet in which calories from fat would be 30 percent or less?

2. Jane is a 35 year old who is 65" tall and weighs 130 pounds. She consumes 54 grams of protein, 150 grams of carbohydrate, and 76 grams of fat per day.

    a. How many calories does Jane consume?

b. What is the percent of calories from carbohydrate in Jane's diet? Does that amount represent an appropriate diet to lower LDL cholesterol based on National Heart, Lung, and Blood Institute recommendations?

Answer these controversy questions:

1. Describe general characteristics of an obesity-promoting environment.

2. How have changes in family structure influenced people's food intakes over the past five decades?

3. What are some of the techniques used to sell foods to youth?

4. What are problems with previous initiatives and attempts by the government to reverse obesity trends by focusing on individual responsibility?

5. How can monetary incentives be used to encourage food producers and consumers toward healthier food and exercise choices?

Study Aids

1. Complete the following table on adult standards for blood pressure, BMI, blood lipids, and risk of CVD.

Blood Pressure

Systolic and diastolic pressure:
120-139/ 80-89 = _____ risk
_____/_____ = healthy

Total Cholesterol

Below _____ mg/dl = healthy
200-239 mg/dl = _____
_____ mg/dl = unhealthy

LDL Cholesterol

Below _____ mg/dl = healthy
130-159 mg/dl = _____
_____ mg/dl = unhealthy

Obesity

Body mass index greater than: _____ = unhealthy

HDL Cholesterol

HDL< _____ mg/dl = unhealthy
≥ 60 mg/dl = _____

Triglycerides (Fasting)

< _____ mg/dl = healthy

2. Use Table 11-9 on page 423 of your textbook to study the dietary guidelines for lowering disease risks.

Sample Test Items

Comprehension Level Items:

1. Nutrition can directly prevent or cure infectious diseases.

   a. true
   b. false

2. Effects of malnutrition on the body's defense systems include:

   a. cells of the immune system are reduced in size.
   b. antibody concentration becomes depressed.
   c. skin becomes thick with more connective tissue.
   d. a and b
   e. b and c

3. A deficiency or a toxicity of even a single nutrient can seriously weaken even a healthy person's immune defense.

    a. true
    b. false

4. Degenerative diseases arise from a mixture of:

    a. genetic predisposition.
    b. personal medical history.
    c. infection.
    d. a and b
    e. a and c

5. Diet is not the only, and perhaps not even the most important, factor in development of CVD.

    a. true
    b. false

6. Which of the following is **not** part of dietary changes recommended to decrease LDL cholesterol?

    a. decrease dietary saturated fat
    b. decrease dietary cholesterol
    c. decrease insoluble fiber
    d. weight reduction

7. To lower LDL and reduce heart disease risk, people in the United States should:

    a. consume no more than 35% of their calories as fat.
    b. consume no more than 20% of their calories as saturated fat.
    c. consume no more than 300 milligrams of cholesterol per day.
    d. a and b
    e. a and c

8. Which of the following correlates directly with a higher risk of heart disease?

    a. high total cholesterol
    b. high LDL cholesterol
    c. high triglycerides
    d. high HDL cholesterol

9. Physical activity and weight loss lower LDL, raise HDL, improve insulin sensitivity, and lower blood pressure.

    a. true
    b. false

10. To help lower blood cholesterol you should:

    a. reduce the amount of saturated fat in the diet.
    b. shun away from eggs altogether.
    c. choose foods high in soluble fiber.
    d. a and b
    e. a and c

11. The twin demons that worsen cardiovascular disease are:

    a.   atherosclerosis.
    b.   hypertension.
    c.   strokes.
    d.   a and b
    e.   b and c

12. Most people who develop hypertension do so in their:

    a.   20s and 30s.
    b.   30s and 40s.
    c.   40s and 50s.
    d.   50s and 60s.

13. The best kind of exercise to reduce hypertension is the prolonged kind, such as walking or jogging, recommended for increasing HDL and lowering LDL.

    a.   true
    b.   false

14. All of the following are likely to be sodium sensitive individuals **except** for:

    a.   those of European or Asian descent.
    b.   those with kidney problems.
    c.   those with diabetes.
    d.   those whose parents have hypertension.

15. For people with hypertension, a reduction of blood pressure usually accompanies a lower salt intake.

    a.   true
    b.   false

16. For people who are overweight and hypertensive, a loss of _____ pounds may significantly lower blood pressure.

    a.   2
    b.   5
    c.   8
    d.   10

17. Which of the following would you do to consume more phytochemicals in your diet?

    a.   eat more fruit
    b.   take dietary supplements
    c.   choose whole rather than refined grains
    d.   a and b
    e.   a and c

18. To minimize risks from carcinogens formed during cooking meats you would:

    a.   bake the food in the oven.
    b.   grill the food.
    c.   pan-fry the food.
    d.   smoke the food.

19. Cruciferous vegetables:

  a. are associated with low cancer rates.
  b. include vegetables such as cauliflower, cabbage, and brussels sprouts.
  c. contain chemicals that act as promoters of cancer.
  d. a and b
  e. b and c

20. The DRI Committee's Tolerable Upper Intake Level for sodium is _____ milligrams per day.

  a. 1200
  b. 1450
  c. 2050
  d. 2300

21. Sterol and stanol esters added to foods such as margarine help to lower blood cholesterol by:

  a. preventing cholesterol from being digested.
  b. removing cholesterol from the food.
  c. reducing the cholesterol in the food.
  d. blocking absorption of cholesterol from the intestine.

22. For middle-aged and older people, moderate alcohol consumption has been reported to raise HDL cholesterol concentrations or reduce the risk of blood clots.

  a. true
  b. false

23. Which of the following is (are) recommended as sources of essential fatty acids that confer benefits on the heart?

  a. flaxseed oil
  b. fish oil supplements
  c. nuts
  d. a and b
  e. a and c

24. Features of the metabolic syndrome include all of the following **except:**

  a. low HDL.
  b. high blood pressure.
  c. low triglycerides.
  d. high fasting blood glucose.

Application Level Items:

25. You are offering help to an AIDS patient. To support his nutrition status you would:

  a. provide three large meals per day.
  b. insist that food be consumed rather than supplements.
  c. discourage the use of liquid meal replacements.
  d. thoroughly cook all foods and practice cleanliness.

26. You are concerned that you may be at risk for developing CVD. Which of the following would indicate that you are at risk?

    a. BMI of 30
    b. your father died of a heart attack at age 53
    c. your HDL is 60 mg/dl
    d. a and b
    e. b and c

27. To increase consumption of fiber you would do all of the following **except**:

    a. consume more dried peas and beans.
    b. eat more fresh fruit.
    c. choose more white breads.
    d. eat more whole-grain foods.

28. You are a 35-year-old male who has just had a physical exam, the results of which are shown below. Which of these results indicate that you are at risk for CVD?

    a. blood cholesterol of 180 mg/dl
    b. body mass index of 22
    c. HDL cholesterol of 35 mg/dl
    d. blood pressure of 110/70

29. To decrease your risk of cancer you would:

    a. decrease your intake of fat.
    b. increase your intake of smoked foods.
    c. eat foods high in fiber.
    d. a and b
    e. a and c

30. Henry is interested in reducing his risk for developing hypertension. He should do all of the following **except**:

    a. lose weight if he is overweight.
    b. moderate alcohol consumption.
    c. decrease his consumption of potassium.
    d. moderately restrict salt intake.

Answers

Summing Up

(1) degenerative; (2) infectious; (3) chronic; (4) immune; (5) food; (6) protein-energy malnutrition; (7) wasting; (8) drug; (9) hospital; (10) independence; (11) degenerative; (12) risk factors; (13) behavioral; (14) obese; (15) third; (16) cancer; (17) diet; (18) medical; (19) obese; (20) physical inactivity; (21) diabetes; (22) LDL; (23) LDL; (24) cholesterol; (25) atherosclerosis; (26) insulin resistance; (27) age; (28) salt; (29) DASH; (30) 20-50; (31) fruits; (32) whole wheat; (33) phytochemicals

Chapter Glossary

Crossword Puzzle: (1A) diastolic pressure; (1D) degenerative diseases; (2) carcinogenesis; (3) systolic pressure; (4) embolism; (5) bioterrorism; (6) AIDS; (7) thrombosis; (8) infectious diseases; (9) aorta
Matching Exercise: (1) h; (2) m; (3) b; (4) k; (5) p; (6) d; (7) s; (8) f; (9) r; (10) a, (11) n; (12) i; (13) g: (14) q; (15) l; (16) c; (17) o; (18) j; (19) e

Exercises

Chapter Study Questions:

1. The antibody concentration and number of cells normally present in secretions of the digestive tract are depressed in malnutrition which allows infectious agents that would normally be barred from the body to enter. Once they are inside the body, the defense mounted against them is weak.

2. Because a common bacteria in food, such as *Salmonella*, can cause a deadly infection in people with compromised immunity, such as individuals with AIDS.

3. You should consider your family's medical history to see which diseases are common and also notice which test results are out of line during your physical examination.

4. When an embolus lodges in an artery of the heart and causes sudden death of part of the heart muscle, a heart attack occurs. A stroke occurs when the embolus lodges in an artery of the brain and kills a portion of brain tissue.

5. Although antioxidant nutrients are found in vegetables and fruits, many other nutrients, such as fiber, folate, and calcium are thought to be protective. Vitamin supplements would not supply all of these.

6. The primary features include central obesity, low HDL and high blood triglycerides, elevated blood pressure, and elevated fasting blood glucose or insulin resistance.

Short Answer Questions:

1. (a) people who restrict their food intake; (b) the very young or old; (c) the hospitalized; (d) the poor

2. (a) environmental; (b) behavioral; (c) social; (d) genetic

3. (a) increasing age; (b) male gender; (c) family history of premature heart disease

4. (a) obesity; (b) atherosclerosis; (c) insulin resistance

5. (a) exposure to a carcinogen; (b) entry of the carcinogen into a cell; (c) initiation; (d) acceleration of other carcinogens called promoters; (e) spreading of cancer cells via blood and lymph; (f) disruption of normal body functions

6. (a) mouth; (b) throat; (c) breast

7. (a) central obesity; (b) abnormal blood lipids; (c) elevated blood pressure; (d) elevated fasting glucose or insulin resistance

8. (a) cabbage; (b) cauliflower; (c) broccoli; (d) Brussels sprouts; (e) greens; (f) rutabagas

9. (a) well-done; (b) grilled; (c) fried; (d) highly browned and broiled; (e) smoked

10. (a) healthy body weight; (b) regular physical activity; (c) moderation in those who use alcohol; (d) diet high in fruits, vegetables, fish and low-fat dairy products, and low in fat; (e) reduction in salt and sodium intake

Problem-Solving:

1. 26% calories from fat; yes

2. (a) 1500 calories; (b) percentage calories from carbohydrate is 40%; no, because carbohydrates should total 50-60% of calories

Controversy Questions:

1. It is an environment which promotes a sedentary lifestyle and over-consumption of high-energy foods.

2. More women have sought careers and entered the work force, which means that they have less time to devote to the traditional activities of homemaking, including less meal planning, grocery shopping, and labor-intensive preparation required for meals of fresh foods. As a result, more prepared or partially-prepared foods are used.

3. The techniques include manipulating children's emotional and physical needs, such as needs for peer acceptance, love, safety, security, desire for independence, to act older, or to develop an identity.

4. They fail to acknowledge societal features that present barriers in the path of those who try to follow health-promoting advice.

5. Taxes could be raised on high-calorie, high-fat, high-salt foods, sugary sodas, candies, and fried snack foods, and the revenues used to subsidize purchase of healthier food choices.

Study Aids

Blood Pressure

Systolic and diastolic pressure:
120-139/ 80-89   = borderline risk
< 120 / < 80   = healthy

Total Cholesterol

Below 200 mg/dl   = healthy
200-239 mg/dl   = borderline
≥ 240 mg/dl   = unhealthy

LDL Cholesterol

Below 100 mg/dl   = healthy
130-159 mg/dl   = borderline
160-189 mg/dl   = unhealthy

Obesity

Body mass index greater than: 30   = unhealthy

HDL Cholesterol

HDL< 40 mg/dl   = unhealthy
≥ 60 mg/dl   = healthy

Triglycerides (Fasting)

< 150 mg/dl   = healthy

Sample Test Items

1. b (p. 394)	9. a (p. 404)	17. e (p. 420)	24. c (p. 405)
2. d (p. 395-396)	10. e (p. 406-407)	18. a (p. 418)	25. d (p. 396)
3. a (p. 396)	11. d (p. 400)	19. d (p. 420)	26. d (p. 401-402)
4. d (p. 394)	12. d (p. 409)	20. d (p. 411)	27. c (p. 406)
5. a (p. 401)	13. a (p. 411)	21. d (p. 407)	28. c (p. 402)
6. c (p. 406-407)	14. a (p. 411)	22. a (p. 407)	29. e (p. 418-419)
7. e (p. 406)	15. a (p. 411)	23. e (p. 406)	30. c (p. 411-412)
8. b (p. 401)	16. d (p. 411)		

Chapter 12 - Food Safety and Food Techology

Chapter Objectives
After completing this chapter, you should be able to:

1. List and prioritize the areas of concern in the U.S. food supply identified by the Food and Drug Administration.

2. Discuss how microbial food poisoning can be prevented and indicate which foods are particularly troublesome.

3. Describe the risks, if any, of the following in foods: natural toxins, environmental contaminants, pesticides, and hormone residues.

4. List five important food processing or preservation techniques and explain the effect they have on the nutrient content of foods.

5. Discuss the regulations concerning food additives and identify the special roles of the major classes of additives.

6. List the arguments for and against the use of organic foods and products of biotechnology (Controversy 12).

Key Concepts
- ✓ Each year in the United States, many millions of people suffer mild to life-threatening symptoms caused by foodborne illness. Though relatively rare, botulism is often deadly when it strikes.

- ✓ Industry employs sound practices to safeguard the commercial food supply from microbial threats. Still, incidents of commercial foodborne illness have caused widespread harm to health.

- ✓ To prevent foodborne illness, always remember that it can happen. Keep hot foods hot, keep cold foods cold, keep raw foods separate, and keep your hands and the kitchen clean.

- ✓ Some foods pose special microbial threats and so require special handling. Raw seafood is especially likely to be contaminated. Honey is unsafe for infants. Almost all types of food poisoning can be prevented by safe food preparation, storage, and cleanliness.

- ✓ Some special food-safety concerns arise when traveling. To avoid foodborne illnesses, remember to boil it, cook it, peel it, or forget it.

- ✓ Natural foods contain natural toxins that can be hazardous if consumed in excess. To avoid poisoning by toxins, eat all foods in moderation, treat chemicals from all sources with respect, and choose a variety of foods.

- ✓ Pesticides can be part of a safe food production process but can also be hazardous if mishandled. The FDA tests for pesticide residues in both domestic and imported foods. Consumers can take steps to minimize their ingestion of pesticide residues in foods.

- ✓ Bovine somatotropin causes cattle to produce more meat and milk on less feed than untreated cattle. The FDA has deemed the practice safe. Antibiotic overuse fosters antibiotic resistance in bacteria, threatening human health. Arsenic drugs are used to promote growth in chickens and other livestock.

- ✓ Persistent environmental contaminants pose a significant, but generally small, threat to the safety of food. An accidental spill can create an extreme hazard. Mercury and other contaminants are of greatest concern during pregnancy, lactation, and childhood.

✓ Some nutrients are lost in food processing. Processing aims to protect food from microbial, oxidative, and enzymatic spoilage.

✓ Modified atmosphere packaging makes many fresh packaged foods available to consumers. MAP foods compare well to fresh foods in terms of nutrient quality. MAP foods may pose a threat of foodborne illness if not properly stored. Consumers can judge freshness of packaged foods by reading the dates on their labels.

✓ Some water-soluble vitamins are destroyed by canning, but many more diffuse into the canning liquid. Fat-soluble vitamins and minerals are not affected by canning, but minerals also leach into canning liquid.

✓ Foods frozen promptly and kept frozen lose few nutrients.

✓ Commercially dried foods retain most of their nutrients, but home-dried foods often sustain dramatic losses.

✓ Extrusion involves heat and destroys nutrients.

✓ The FDA regulates the use of intentional additives. Additives must be safe, effective, and measurable in the final product. Additives on the GRAS list are assumed to be safe because they have long been used. Additives used must have wide margins of safety.

✓ Microbial food spoilage can be prevented by antimicrobial additives. Of these, sugar and salt have a long history of use. Nitrites added to meats have been associated with cancer in laboratory animals.

✓ Antioxidants prevent oxidative changes in foods that would lead to unacceptable discoloration and texture changes in the food. Ingestion of the antioxidant sulfites can cause problems for some people; BHT may offer antioxidant effects in the body.

✓ The addition of artificial colors is tightly controlled.

✓ Among flavorings added to foods, the flavor enhancer MSG causes reactions in people with sensitivities to it.

✓ Incidental additives are substances that get into food during processing. They are well regulated, and most present no hazard.

✓ Nutrients are added to foods to enrich or to fortify them. These additives do not necessarily make the foods nutritious, only rich in the vitamins and minerals that have been added.

Summing Up

The (1)_____ is the major agency charged with monitoring the U.S. food supply. According to the FDA, the food hazard which has the potential to cause the greatest harm to people is (2)_____ foodborne illness.

The term *foodborne illness* refers either to foodborne infection or to food (3)_____. The symptoms of one neurotoxin stand out as severe and commonly fatal—those of (4)_____. The botulinum toxin is destroyed by (5)_____ so canned foods that have been boiled for ten minutes are generally safe from this threat.

Food can provide ideal conditions for (6)_____ to thrive or to produce toxins. Disease-causing

bacteria require warmth, (7)_____, and nutrients. To defeat them, people who prepare food should

keep hot food hot, keep cold food cold, keep raw foods separate, and keep your hands and the

(8)_____ clean. Foods that are high in moisture and nutrients and those that are chopped or

(9)_____ are especially favorable hosts for microbial growth. You cannot rely on your senses of smell

and sight alone to warn you of hazards because most contamination is not detectable by (10)_____,

taste, or appearance. Raw, unpasteurized (11)_____ and raw (12)_____ are especially likely

to be contaminated.

A food (13)_____ is any substance occurring in food by accident, and the potential harmfulness

depends in part on the extent to which it lingers in the environment or in the human body. In general, the threat from

environmental contaminants is small because the (14)_____ monitors the presence of contaminants in

foods. Risks to health from pesticide exposure are probably small for healthy adults, but (15)_____

may be more at risk for pesticide poisoning.

Canning, (16)_____, drying, extrusion and modified atmosphere processing are important

techniques of processing foods. In general, frozen foods' nutrient contents are similar to those of

(17)_____ foods and losses are minimal, although (18)_____ destroys some water-

soluble vitamins.

Manufacturers use food (19)_____ to give foods desirable characteristics, including color,

flavor, (20)_____, stability, enhanced nutrient composition, or resistance to spoilage. The use of

additives is regulated by the FDA and additives must be safe, (21)_____ and detected and measured

in the final food product. Additives included on the (22)_____ list are assumed to be safe because

they have been used for a long time and their use entailed no known hazards.

Preservatives known as (23)_____ agents are used to protect foods from growth of microbes.

The two best known and most widely used agents include sugar and (24)_____; both work by

withdrawing (25)_____ from the food. Another group of antimicrobial agents, the

(26)_____, are added to foods to preserve their color, to enhance their flavor, and to protect against

bacterial growth, especially the deadly (27)_____ bacterium.

Some foods go bad by undergoing changes in color and flavor caused by exposure to (28)_____ in the air. Preservatives known as (29)_____ protect food from this type of spoilage. Vitamin E and vitamin (30)_____ are used to prevent oxidation, as well as sulfites, BHA, and (31)_____.

Another class of additives is nutrients added to improve or maintain the (32)_____ value of foods. Included among these are the nutrients added to (33)_____ grains to enrich them, the (34)_____ added to salt, vitamins A and (35)_____ added to dairy products, and the nutrients added to fortify breakfast cereals.

Chapter Glossary

Crossword Puzzle:

Across:	*Down:*
2. an often-fatal illness of cattle affecting the nerves and brain; also called *bovine spongiform encephalopathy* (BSE)	1. an infective agent consisting of an unusually folded protein that disrupts normal cell functioning, causing disease
4. a hormone (somatotropin) that promotes growth and that is produced naturally in the pituitary gland of the brain	3. poisons that act upon mucous membranes such as those of the digestive tract
7. the treatment of milk with heat sufficient to kill certain pathogens commonly transmitted through milk; not a sterilization process	5. a systematic plan to identify and correct potential microbial hazards in the manufacturing, distribution, and commercial use of food products (first 2 of 5 words in the answer)
9. any of a number of mineral ions such as mercury and lead	6. a systematic plan to identify and correct potential microbial hazards in the manufacturing, distribution, and commercial use of food products (last 3 of 5 words in the answer)
12. the maximum amount of a residue permitted in a food when a pesticide is used according to label directions	8. the accumulation of a contaminant in the tissues of living things at higher and higher concentrations along the food chain
13. poisons that act upon the cells of the nervous system	9. a state of danger; used to refer to any circumstance in which harm is possible under normal conditions of use
14. an often-fatal food poisoning caused by botulinum toxin, a toxin produced by bacteria that grow without oxygen in nonacidic canned foods	10. minute organisms too small to observe without a microscope, including bacteria, viruses, and others
15. growth hormone of cattle which can be produced for agricultural use by genetic engineering	11. the science that manipulates biological systems or organisms to modify their products or components or create new products
16. illness transmitted to human beings through food and water	

Word Bank:		
bioaccumulation	growth hormone	microbes
biotechnology	hazard	neurotoxins
botulism	Hazard Analysis Critical Control Point	pasteurization
bovine somatotropin	heavy metal	prion
entertoxins	mad cow disease	tolerance limit
foodborne illness		

Matching Exercise:

_____ 1. incidental additives

_____ 2. canning

_____ 3. modified atmosphere packaging

_____ 4. high-temperature-short-time principle

_____ 5. contaminant

_____ 6. MSG symptom complex

_____ 7. freezing

_____ 8. extrusion

_____ 9. GRAS list

_____ 10. drying

_____ 11. pesticides

_____ 12. additives

_____ 13. margin of safety

_____ 14. toxicity

_____ 15. ultrahigh temperature

_____ 16. organic foods

_____ 17. cross-contamination

a. the acute, temporary, and self-limiting reactions experienced by sensitive people upon ingesting a large dose of MSG

b. preservation by removing sufficient water from food to inhibit microbial growth

c. substances that are added to foods, but are not normally consumed by themselves as foods

d. the ability of a substance to harm living organisms

e. preservation of a perishable food by packaging it in a gas-impermeable container from which air has been removed, or to which another gas mixture has been added

f. a process by which the form of food is changed, such as changing corn to corn chips; not a preservation measure

g. in reference to food additives, a zone between the concentration normally used and that at which a hazard exists

h. every 10º C (18º F) rise in processing temperature brings about an approximately tenfold increase in microbial destruction, while only doubling nutrient losses

i. substances that can get into food not through intentional introduction but as a result of contact with food during growing, processing, packaging, storing, or some other stage before the food is consumed

j. a list of food additives, established by the FDA, long in use and believed safe

k. a process of sterilizing food by exposing it for a short time to temperatures above those normally used in processing

l. preservation by lowering food temperature to the point at which life processes cease

m. chemicals used to control insects, diseases, weeds, fungi, and other pests on crops and around animals

n. preservation by killing all microorganisms present in food and sealing out air

o. any substance occurring in food by accident; not a normal food constituent

p. the contamination of a food through exposure to utensils, hands, or other surfaces that were previously in contact with a contaminated food

q. products grown and processed without the use of synthetic chemicals such as pesticides, herbicides, fertilizers, and preservatives and without genetic engineering or irradiation

Exercises

Answer these chapter study questions:

1. Why have food colors been more heavily criticized than almost any other group of additives?

2. Identify the steps which the FDA has taken to protect people who are allergic to sulfites.

3. Why are some foods more encouraging of microbial growth than others?

4. Explain what is meant by the statement that food processing involves a trade-off.

5. Why are vitamin C losses especially likely during the freezing process?

6. What is the Hazard Analysis Critical Control Point (HACCP) plan?

Complete these short answer questions:

1. Manufacturers use food additives to give foods desirable characteristics related to:

   a.
   b.
   c.
   d.
   e.
   f.

2. The best known and most widely used antimicrobial agents are:

   a.
   b.

3. Nitrites are added to foods to:

   a.
   b.
   c.

4. Four food products which commonly contain nutrient additives are:

   a.
   b.
   c.
   d.

5. Incidental food additives are substances that find their way into food as the result of some phase of:

   a.
   b.
   c.
   d.

6. Warning signs of botulism include:

   a.
   b.
   c.
   d.

7. Food preservation or processing techniques include:

   a.
   b.
   c.
   d.
   e.

8. The canning process is based on:

   a.
   b.

9. Advantages of drying as a food processing method include:

   a.
   b.
   c.

10. Four things which should be done in the kitchen in order to prevent food poisoning include:

    a.
    b.
    c.
    d.

Solve these problems:

For questions 1-4, match the types of food additives on the left with examples of the additives on the right.

_____ 1. antimicrobial agent	a.	MSG
_____ 2. antioxidant	b.	sulfites
_____ 3. artificial flavor	c.	iodine
_____ 4. nutrient additive	d.	nitrites

For questions 5-8, for each food listed on the left, identify proper storage and/or cooking techniques to preserve the nutrient content, listed on the right.

_____ 5. milk	a.	chill immediately after picking and keep cold until used
_____ 6. apples	b.	cut after washing; add to water that is vigorously boiling
_____ 7. cut orange	c.	cover with airtight wrapper and store in refrigerator
_____ 8. potatoes	d.	store in cardboard or opaque plastic containers to protect riboflavin

Answer these controversy questions:

1. Which specific technologies cannot be used in the production of organic foods?

2. Would a consumer be able to detect a genetically modified food product in the grocery store? Why or why not?

3. What are plant pesticides and what are their limitations?

4. Are organic foods nutritionally superior to traditional foods? If so, in what way?

5. What are some of the potential health risks associated with organic foods?

Study Aids

Study Table 12-5 on page 444 of your textbook to determine the safe refrigeration storage times for various foods, and determine ways to reduce pesticide residue intake by studying Table 12-11 on page 454 of your textbook.

Sample Test Items

Comprehension Level Items:

1. Which of the following appears as the number one area of concern on the Food and Drug Administration's list of concerns regarding our food supply?

    a. pesticide residues
    b. foodborne illness
    c. intentional food additives
    d. naturally occurring toxicants

2. Substances that find their way into food as a result of production, processing, storage, or packaging are called
    _____ additives.

    a. incidental
    b. contaminant
    c. intentional
    d. direct

3. FDA's authority over additives hinges primarily on their:

    a. appropriate uses.
    b. effectiveness.
    c. safety.
    d. a and b
    e. b and c

4. If FDA approves an additive's use, that means the manufacturer can:

    a. use it only in amounts specified by the FDA.
    b. add it to only those foods delineated by the FDA.
    c. add it in any amount to any food.
    d. a and b
    e. b and c

5. Which of the following is(are) true concerning the GRAS list?

    a. Some 700 substances have been placed on the list.
    b. Substances on the GRAS list can be re-evaluated.
    c. Substances on the GRAS list cannot be changed.
    d. a and b
    e. b and c

6. MSG has been deemed safe for adults, but is kept out of foods for infants.

    a. true
    b. false

7. Salt and sugar work as antimicrobial agents by:

    a. removing oxygen from foods.
    b. withdrawing nutrients used as food by bacteria.
    c. withdrawing water from food.
    d. b and c
    e. a and b

8. Preservatives known as _____ protect food from undergoing changes by exposure to air.

    a. glutamates
    b. tartrazines
    c. antimicrobials
    d. antioxidants

9. To reduce pesticide residues in foods you would:

   a. trim the fat from meat.
   b. consume waxed vegetables without removing the wax.
   c. wash fresh produce in water.
   d. a and b
   e. a and c

10. No food additives are permanently approved by the FDA.

   a. true
   b. false

11. To prevent bacterial growth, you should refrigerate foods immediately after serving a meal and definitely before _____ has(have) passed.

   a. 30 minutes
   b. 1 hour
   c. 2 hours
   d. 3 hours

12. Which of the following would be the safest food to pack for a picnic?

   a. carrot sticks
   b. pasta salad
   c. chicken salad
   d. Swiss cheese

13. A processing technique that preserves food by lowering food temperature to the point at which life processes cease is called:

   a. canning.
   b. freezing.
   c. drying.
   d. extrusion.

14. Which of the following is a special enemy of riboflavin?

   a. heat
   b. the canning process
   c. light
   d. ascorbic acid oxidase

15. You can use your senses of smell and sight alone to warn you of hazards and food contamination.

   a. true
   b. false

16. A mineral added when foods are canned is:

   a. potassium.
   b. iron.
   c. calcium.
   d. sodium.

17. Sulfites are highly destructive of:

    a. vitamin C.
    b. riboflavin.
    c. iron.
    d. thiamin.

18. Wooden cutting boards do not support microbial growth and therefore require no treatment.

    a. true
    b. false

19. To prevent traveler's diarrhea you would:

    a. wash hands before eating.
    b. drink only bottled beverages.
    c. wash fresh fruits under running water.
    d. a and b
    e. b and c

20. Which of the following appears last on the FDA's list of concerns regarding the U.S. food supply?

    a. residues in foods
    b. genetic modification of foods
    c. natural toxins in foods
    d. intentional food additives

21. Foodborne illnesses can be life-threatening and are increasingly unresponsive to standard antibiotics.

    a. true
    b. false

22. Which of the following is(are) characteristic of a safe hamburger?

    a. one cooked until medium well done
    b. one that has reached an internal temperature of 160° F
    c. one in which the juices run clear
    d. a and b
    e. b and c

23. To decrease foodborne illness, you would rinse all produce thoroughly under running water for at least 10 seconds.

    a. true
    b. false

24. The FDA and EPA advise all pregnant women against eating king mackerel, swordfish and shark because they are high in:

    a. lead.
    b. mercury.
    c. selenium.
    d. arsenic.

Application Level Items:

25. Which of the following food products would be acceptable to purchase from the grocery store?

    a. a jar of jelly with a broken seal
    b. a solidly frozen package of spinach
    c. a can of tomatoes that is bulging
    d. a torn package of frozen green peas

26. Which of the following foods would be most susceptible to bacterial contamination?

    a. pork chop
    b. chicken breast
    c. meatloaf
    d. veal cutlet

27. Which of the following foods would be most appropriate to carry on a picnic?

    a. pimento cheese spread
    b. meatballs
    c. egg salad
    d. canned cheese spread

28. To prevent food poisoning you would:

    a. transfer leftovers to deep containers before refrigerating.
    b. refrigerate perishables immediately when you get home.
    c. thaw poultry in the refrigerator.
    d. a and b
    e. b and c

29. The FDA approved the use of Aspartame as a sweetener several years ago. This means that:

    a. it can be added to only certain approved food products.
    b. it can be added in unlimited quantities to foods.
    c. it can be used forever in food products.
    d. it can be placed on the GRAS list.

30. You are interested in using a preservative which will protect a food from microbes and prevent the food from becoming hazardous to health. Which of the following would you use?

    a. ascorbate
    b. sulfites
    c. tartrazine
    d. nitrites

Answers

Summing Up

(1) Food and Drug Administration; (2) microbial; (3) intoxication; (4) botulism; (5) heat; (6) bacteria; (7) moisture; (8) kitchen; (9) ground; (10) odor; (11) eggs, (12) seafood; (13) contaminant; (14) FDA; (15) children; (16) freezing; (17) fresh; (18) canning; (19) additives; (20) texture; (21) effective; (22) GRAS; (23) antimicrobial; (24) salt; (25) water; (26) nitrites; (27) botulinum; (28) oxygen; (29) antioxidants; (30) C; (31) BHT; (32) nutritional; (33) refined; (34) iodine; (35) D

Chapter Glossary

Crossword Puzzle: (1) prion; (2) mad cow disease; (3) entertoxins; (4) growth hormone; (5) Hazard Analysis; (6) Critical Control Point; (7) pasteurization; (8) bioaccumulation; (9 A) heavy metal; (9 D) hazard; (10) microbes; (11) biotechnology; (12) tolerance limit; (13) neurotoxins; (14) botulism; (15) bovine somatotropin; (16) foodborne illness

Matching Exercise: (1) i; (2) n; (3) e; (4) h; (5) o; (6) a; (7) l; (8) f; (9) j; (10) b; (11) m; (12) c; (13) g; (14) d; (15) k; (16) q; (17) p

Exercises

Chapter Study Questions:

1. Because they are dispensable and are used to make foods look pretty. In contrast, other food additives, such as preservatives, make foods safe.

2. It prohibits sulfite use on food intended to be consumed raw, with the exception of grapes; it requires that sulfite-containing foods and drugs list the additives on their labels to warn that sulfites are present.

3. Because the ideal conditions for bacteria to thrive include warmth, moisture, and nutrients; foods that are high in moisture and nutrients and those that are chopped or ground are especially favorable hosts for microbial growth.

4. Food processing makes food safer, gives it a longer usable lifetime than fresh food, or cuts preparation time; however, the cost is the loss of some vitamins and minerals.

5. Vitamin C losses occur whenever tissues are broken and exposed to the oxygen in the air. In the freezing process, vitamin C losses are likely to occur during the steps taken in preparation for freezing, including blanching, washing, trimming, or grinding.

6. HACCP is a method of preventing foodborne illnesses at their source. Each producer, packer, distributor, and transporter of food identifies critical control points in its procedures where the risk of food contamination is high, and then devises and implements ways of minimizing contamination.

Short Answer Questions:

1. (a) color; (b) flavor; (c) texture; (d) stability; (e) resistance to spoilage; (f) enhanced nutrient composition

2. (a) sugar; (b) salt

3. (a) preserve their color; (b) enhance their flavor by inhibiting rancidity; (c) protect against bacterial growth

4. (a) refined grains; (b) salt; (c) dairy products; (d) breakfast cereals

5. (a) production; (b) processing; (c) storage; (d) packaging

6. (a) double vision; (b) weak muscles; (c) difficulty swallowing; (d) difficulty breathing

7. (a) canning; (b) freezing; (c) drying; (d) extrusion; (e) modified atmosphere packaging

8. (a) time; (b) temperature

9. (a) it eliminates microbial spoilage; (b) it reduces the weight and volume of foods; (c) it does not cause major nutrient losses

10. (a) keep hot foods hot; (b) keep cold foods cold; (c) keep your hands and the kitchen clean; (d) keep raw food separate

Problem-Solving:

(1) d; (2) b; (3) a; (4) c; (5) d; (6) a; (7) c; (8) b

Controversy Questions:

1. Organic foods must be free of ingredients produced by way of specific technologies, such as irradiation and biotechnology, including genetic engineering.

2. No, because no label is required for genetically modified products.

3. Plant pesticides are a group of pesticides made within the tissues of plants themselves; limitations include a lack of control of pests not killed by the plant pesticide, the high cost of producing the transgenic plants, and the development of resistance to the pesticide among the target population of insects.

4. Some findings suggest that organic foods have slightly greater vitamin C content, lower nitrate, and somewhat improved protein quality but the differences were small and not significant to human health.

5. The use of improperly composted animal manure fertilizer may expose consumers to *E. coli*; unpasteurized organic juices, milk and cheese may also constitute microbial hazards because pasteurization is required to kill disease-causing microorganisms in these foods; also, organic foods contain no preservatives and tend to spoil faster than other foods.

Sample Test Items

1. b (p. 434)	11. c (p. 440)	21. a (p. 435)
2. a (p. 466)	12. a (p. 447)	22. e (p. 445)
3. e (p. 462)	13. b (p. 461)	23. a (p. 446)
4. d (p. 463)	14. c (p. 469)	24. b (p. 458)
5. d (p. 463)	15. b (p. 440)	25. b (p. 440)
6. a (p. 466)	16. d (p. 461)	26. c (p. 444)
7. c (p. 464)	17. d (p. 465)	27. d (p. 447-448)
8. d (p. 465)	18. b (p. 442)	28. e (p. 440)
9. e (p. 454)	19. d (p. 448)	29. a (p. 463)
10. a (p. 463)	20. b (p. 434)	30. d (p. 464)

Chapter 13 - Life Cycle Nutrition
Mother and Infant

Chapter Objectives

After completing this chapter, you should be able to:

1. Describe how maternal nutrition before and during pregnancy affects both the development of the fetus and growth of the infant after birth.

2. Discuss maternal physiological adjustments that occur during pregnancy and explain how they influence energy and other nutrient requirements.

3. Explain why abstinence from smoking and drugs, avoiding dieting, and moderation in the use of caffeine are recommended during pregnancy.

4. Explain the effects of alcohol on the development of the fetus and describe the condition known as fetal alcohol syndrome.

5. List the benefits of breastfeeding an infant and indicate the changes a lactating woman needs to make in her diet to promote breastfeeding success.

6. Describe the circumstances under which breastfeeding is not recommended and explain what alternatives exist for proper infant nutrition.

7. Plan a timetable for feeding foods to an infant from birth to 12 months of age.

8. Describe what is currently known about relationships between childhood obesity and early chronic diseases (Controversy 13).

Key Concepts

✓ Adequate nutrition before pregnancy establishes physical readiness and nutrient stores to support fetal growth. Both underweight and overweight women should strive for appropriate body weights before pregnancy. Newborns who weigh less that 5 ½ pounds face greater health risks than normal-weight babies. The healthy development of the placenta depends on adequate nutrition before pregnancy.

✓ Implantation, fetal development, and early critical periods depend on maternal nutrition before and during pregnancy.

✓ Pregnancy brings physiological adjustments that demand increased intakes of energy and nutrients. A balanced diet that includes more nutrient dense foods from the five food groups can help to meet these needs.

✓ Due to their key roles in cell reproduction, folate and vitamin B_{12} are needed in large amounts during pregnancy. Folate plays an important role in preventing neural tube defects.

✓ All pregnant women, but especially those who are less than 25 years of age, need to pay special attention to ensure adequate calcium intakes. A daily iron supplement is recommended for all pregnant women during the second and third trimesters.

- ✓ Women most likely to benefit from multivitamin-mineral supplements during pregnancy include those who do not eat adequately, those carrying twins or triplets, and those who smoke cigarettes or are alcohol or drug abusers.

- ✓ Food assistance programs such as WIC can provide nutritious food for pregnant women of limited financial means.

- ✓ Weight gain is essential for a healthy pregnancy. A woman's prepregnancy BMI, her own nutrient needs, and the number of fetuses she is carrying help to determine appropriate weight gain.

- ✓ Physically fit women can continue to be physically active throughout pregnancy. Pregnant women should be cautious in their choice of activities.

- ✓ Of all the population groups, pregnant teenage girls have the highest nutrient needs and an increased likelihood of having problem pregnancies.

- ✓ Food cravings usually do not reflect physiological needs, and some may interfere with nutrition. Nausea arises from normal hormonal changes of pregnancy.

- ✓ Abstaining from smoking and other drugs, limiting intake of foods known to contain unsafe levels of contaminants such as mercury, taking precautions against foodborne illness, avoiding large doses of nutrients, refraining from dieting, using artificial sweeteners in moderation, and limiting caffeine use are recommended during pregnancy.

- ✓ Alcohol limits oxygen delivery to the fetus, slows cell division, and reduces the number of cells organs produce. Alcoholic beverages must bear warnings to pregnant women.

- ✓ The birth defects of fetal alcohol syndrome arise from severe damage to the fetus caused by alcohol. Lesser conditions, ARND and ARBD, may be harder to diagnose but also rob the child of a normal life.

- ✓ Abstinence from alcohol is critical to prevent irreversible damage to the fetus.

- ✓ Gestational diabetes and preeclampsia are common medical problems associated with pregnancy. These should be managed to minimize associated risks.

- ✓ The lactating woman needs extra fluid and enough energy and nutrients to make sufficient milk each day. Malnutrition most often diminishes the quantity of the milk produced without altering quality. Lactation may facilitate loss of the extra fat gained during pregnancy.

- ✓ Breastfeeding is not advised if the mother's milk is contaminated with alcohol, drugs or environmental pollutants. Most ordinary infections such as colds have no effect on breastfeeding. Where safe alternatives are available, HIV-infected women should not breastfeed their infants.

- ✓ Infants' rapid growth and development depend on adequate nutrient supplies, including water from breast milk or formula.

- ✓ Breast milk is the ideal food for infants because it provides the needed nutrients in the right proportions and protective factors as well.

- ✓ Infant formulas are designed to resemble breast milk and must meet an AAP standard for nutrient composition. Special formulas are available for premature infants, allergic infants, and others. Formula should be replaced with milk only after the baby's first birthday.

- ✓ Solid food additions to an infant's diet should begin at about six months and should be governed by the infant's nutrient needs and readiness to eat. By one year, the baby should be receiving foods from all food groups.

✓ The early feeding of the infant lays the foundation for lifelong eating habits. It is desirable to foster preferences that will support normal development and health throughout life.

Summing Up

Before she becomes pregnant, a woman must establish eating habits that will optimally

(1)_____ both the growing fetus and herself and strive for an appropriate body

(2)_____. An underweight woman who fails to gain adequately during pregnancy is most likely to

bear a baby with a dangerously low (3)_____. Infant birthweight is the most potent single indicator

of an infant's future (4)_____ status. Prepregnancy nutrition determines whether the woman will be

able to grow a healthy (5)_____, which is the only source of (6)_____ available to the

fetus.

During the two weeks following fertilization, minimal growth takes place but it is a (7)_____

period developmentally. At eight weeks, the fetus has a complete central nervous system, a beating

(8)_____, and a fully formed (9)_____ system. The gestation period, which lasts

approximately (10)_____ weeks, ends with the (11)_____ of the infant.

During the second trimester, the pregnant woman needs (12)_____ extra calories per day above

the allowance for nonpregnant women. In addition, about (13)_____ more grams of protein are

required. Pregnant women need ample amounts of (14)_____-rich foods to spare their protein and

fuel the fetal brain. Vitamins needed in large amounts during pregnancy, because of their roles in cell reproduction,

include (15)_____ and vitamin B_{12}. Calcium, (16)_____, and magnesium are the

minerals in greatest demand during pregnancy. Although the body conserves (17)_____ even more

than usual during pregnancy, an iron supplement is recommended during the second and third trimesters. The ideal

pattern of weight gain is thought to be 3 ½ pounds during the first (18)_____ months and a

(19)_____ per week thereafter.

The energy cost of producing milk is about (20)_____ calories a day. The nursing mother

should drink at least (21)_____ cups of liquids each day to prevent dehydration. Breastfeeding is

inadvisable if milk is contaminated with (22)_____, drugs, or environmental pollutants.

One of the most important nutrients for infants, as for everyone, is (23)_____ and

(24)_____ milk excels as a source of nutrients for the young infant. With the exception of

(25)_____, breast milk provides all the nutrients a healthy infant needs for the first

(26)_____ months of life.

   Solid food additions to a baby's diet should begin at about (27)_____ months and should be governed by the baby's nutrient needs, physical (28)_____ to handle different forms of foods, and the need to detect allergic reactions. The early feeding of an infant should foster preferences that will support normal (29)_____ and health throughout life and, most importantly, support normal (30)_____ as the child grows.

Chapter Glossary

Matching Exercise:	
_____ 1. critical period	a. the stage of human gestation from the third to eighth week after conception
_____ 2. gestation	b. a system of scoring an infant's physical condition right after birth
_____ 3. amniotic sac	c. the period of about 40 weeks (three trimesters) from conception to birth; the term of a pregnancy
_____ 4. fetus	d. the stage of human gestation from eight weeks after conception until birth of an infant
_____ 5. ovum	e. a birthweight of less than 5 ½ pounds (2,500 grams); used as a predictor of probable health problems in the newborn and as a probable indicator of poor nutrition status of the mother before and/or during pregnancy
_____ 6. embryo	f. the womb, the muscular organ within which the infant develops before birth
_____ 7. zygote	g. a finite period during development in which certain events may occur that will have irreversible effects on later developmental stages
_____ 8. uterus	h. the stage of development, during the first two weeks after conception, in which the fertilized egg embeds itself in the wall of the uterus and begins to develop
_____ 9. placenta	i. the term that describes the product of the union of ovum and sperm during the first two weeks after fertilization
_____ 10. Apgar score	j. the organ of pregnancy in which maternal and fetal blood circulate in close proximity and exchange nutrients and oxygen and wastes
_____ 11. implantation	k. the egg, produced by the mother, that unites with a sperm from the father to produce a new individual
_____ 12. low birthweight	l. the "bag of waters" in the uterus in which the fetus floats
_____ 13. Special Supplemental Food Program for Women, Infants, and Children (WIC)	m. malformations in the skeletal and organ systems associated with prenatal alcohol exposure
_____ 14. environmental tobacco smoke	n. the combination of exhaled smoke (mainstream smoke) and smoke from lighted cigarettes, pipes, or cigars (sidestream smoke) that enters the air and may be inhaled by other people
_____ 15. alcohol-related birth defects (ARBD)	o. a USDA program offering low-income pregnant women or those with infants coupons redeemable for specific foods to supply the nutrients deemed most needed for growth and development
_____ 16. trimester	p. a serious foodborne infection caused by a bacterium found in soil and water that can cause severe brain infection or death in a fetus or newborn
_____ 17. listeriosis	q. a period representing gestation; about 13-14 weeks

Crossword Puzzle:

Word Bank:

alpha-lactalbumin	edema	milk anemia
anencephaly	fetal alcohol syndrome	neural tube defects
ARND	gestational diabetes	preeclampsia
cesarean section	lactation	spina bifida
colostrum	lactoferrin	

Crossword Puzzle Across:	Crossword Puzzle Down:
4. abnormal glucose tolerance appearing during pregnancy 5. the chief protein in human breast milk 8. behavioral, cognitive, or central nervous system abnormalities associated with prenatal alcohol exposure 9. an uncommon and always fatal neural tube defect in which the brain fails to form 11. a group of nervous system abnormalities caused by interruption of the normal early development of the neural tube 12. iron-deficiency anemia caused by drinking so much milk that iron-rich foods are displaced from the diet 14. a milklike secretion from the breasts during the first day or so after delivery before milk appears	1. the cluster of symptoms seen in an infant or child whose mother consumed excessive alcohol during her pregnancy 2. a potentially dangerous condition during pregnancy characterized by edema, hypertension, and protein in the urine 3. one of the most common types of neural tube defects in which gaps in the bones of the spine often allow the spinal cord to bulge and protrude through the gaps, resulting in a number of motor and other impairments 6. production and secretion of breast milk for the purpose of nourishing an infant 7. surgical childbirth, in which the infant is taken through an incision in the woman's abdomen 10. a factor in breast milk that binds iron and keeps it from supporting the growth of the infant's intestinal bacteria 13. accumulation of fluid in the tissues

Exercises

Answer these chapter study questions:

1. Why is an appropriate prepregnancy weight important and how does it relate to pregnancy outcome?

2. Do all women need to add protein-rich foods to their diets in order to obtain the additional daily protein recommended during pregnancy? Why or why not?

3. How does a woman's body conserve iron during pregnancy? Does this mean that an iron supplement should not be taken?

4. What effect does nutritional deprivation of the mother have on breastfeeding?

5. Discuss ways to prevent infant obesity and to encourage eating habits that will support continued normal weight as the child grows.

6. Why have the FDA and EPA issued an advisory to pregnant women against eating large ocean fish, such as king mackarel and sharks?

Complete these short answer questions:

1. Factors associated with low-birthweight infants include:

   a.
   b.
   c.
   d.
   e.

2. The effects of malnutrition during critical periods of pregnancy are seen in:

   a.
   b.
   c.

3. The pregnant woman's need for folate and vitamin B_{12} increase due to:

   a.
   b.

4. These minerals, involved in the normal development of teeth and bones, are in great demand during pregnancy:

   a.
   b.
   c.

5. Prenatal supplements typically provide more of these nutrients than regular supplements:

   a.
   b.
   c.

6. Components of weight gained by the pregnant woman include these lean tissues:

   a.
   b.
   c.
   d.
   e.

7. The nausea of morning sickness can be alleviated by starting the day with:

   a.
   b.

8. An electrolyte formula designed for infants should be provided to an infant on infant formula or breast milk under the following conditions:

   a.
   b.

9. The addition of foods to a baby's diet should be governed by these three conditions:

   a.
   b.
   c.

10. Normal dental development in a baby should be promoted by:

   a.
   b.
   c.
   d.

Solve these problems:

1. How many calories and grams of carbohydrate and protein should be provided in a diet for a 30-year-old pregnant woman who weighs 130 pounds and is at her appropriate weight for height? She is in her second trimester of pregnancy. Her nonpregnant caloric needs are 2000 calories per day.

For questions 2-6, match the nutrients needed in extra amounts during pregnancy, listed on the right, with some of their best food sources, listed on the left.

_____2.	dried fruits	a. folate
_____3.	dark, green leafy vegetables	b. calcium
_____4.	eggs	c. iron
_____5.	milk	d. zinc
_____6.	shellfish	e. vitamin B_{12}

For questions 7-12, identify the correct sequence for introducing the following foods into an infant's diet, by placing an *"a"* by the first food which should be introduced, a *"b"* by the second food to be introduced, etc.

_____ 7.   pureed vegetables
_____ 8.   pieces of soft, cooked vegetables from the table
_____ 9.   yogurt
_____10.   iron-fortified rice cereal
_____11.   mashed fruits
_____12.   finely-chopped meat

Answer these controversy questions:

1. Describe the typical characteristics of children most likely to be obese.

2. What are common characteristics associated with the diagnosis of type 2 diabetes in children?

3. What are characteristics of physically active children versus sedentary children?

4. What are appropriate dietary guidelines for children found to have high blood lipids?

5. What are recommendations for controlling hypertension in children?

Study Aids

1. Complete the following table by filling in the blanks.

   ### First Foods for the Infant

Age (months)	Addition
0-4	Breast milk or (a)_____
4-6	Iron fortified (b)_____ mixed with breast milk, formula, or water
6-8	Begin plain baby food (c)_____, mashed (d)_____ and fruits
8-10	Gradually begin yogurt, finely cut (e)_____, fish, casseroles, cheese, (f)_____, and legumes

2. Use the following tables in your textbook as study aids: study Table 13-2 on page 487 to identify daily food choices for pregnant and lactating women; review Table 13-4 on page 491 for the recommended weight gains for pregnancy; and study Table 13-8 on page 507 to compare and contrast the composition of human milk versus infant formula.

Sample Test Items

Comprehension Level Items:

1. Which of the following is the major factor in low birthweight?

   a. heredity
   b. smoking
   c. drug use
   d. poor nutrition

2. The muscular organ within which the infant develops before birth is called the

   a. placenta.
   b. uterus.
   c. ovum.
   d. zygote.

3. The stage of human gestation from the third to eighth week after conception is called a(an):

   a. ovum.
   b. zygote.
   c. embryo.
   d. fetus.

4. The recommended protein intake during pregnancy is _____ grams per day higher than for nonpregnant women.

   a. 25
   b. 50
   c. 60
   d. 100

5. Folic acid, the synthetic form of folate, is better absorbed than the naturally occurring folate in foods.

   a. true
   b. false

6. A daily iron supplement containing _____milligrams is recommended during the second and third trimesters of pregnancy.

   a. 10
   b. 20
   c. 30
   d. 40

7. A sudden large weight gain during pregnancy is a signal that may indicate:

   a. that the woman is consuming too many calories.
   b. that the baby will be too large.
   c. the onset of preeclampsia.
   d. a and b
   e. b and c

8. Which of the following practices should be avoided during pregnancy?

   a. exercise
   b. dieting
   c. consumption of caffeine
   d. a and b
   e. b and c

9. A strategy which has been found effective in alleviating the nausea of morning sickness is to start the day with:

   a. a few sips of a carbonated beverage.
   b. a food high in fat.
   c. a salty snack food.
   d. a and b
   e. a and c

10. Breast milk volume depends on:

    a. the size of the mother's breasts.
    b. how much fluid the mother consumes.
    c. how much milk the baby demands.
    d. a and b
    e. b and c

11. Characteristics of breast milk include all of the following **except**:

    a. its carbohydrate is in the form of lactose.
    b. it is high in sodium.
    c. its iron is highly absorbable.
    d. its zinc is absorbed better than from cow's milk.

12. A baby less than one year of age who has been on breast milk should be weaned onto:

    a. whole milk.
    b. skim milk.
    c. low-fat milk.
    d. infant formula.

13. After 4-6 months, an exclusively breastfed baby needs:

    a. vitamin C.
    b. vitamin D.
    c. iron.
    d. vitamin A.

14. Food aversions and cravings that arise during pregnancy are usually due to:

    a. changes in taste.
    b. physiological needs.
    c. changes in smell sensitivities.
    d. a and b
    e. a and c

15. Birth defects have been observed in the children of some women who drank _____ ounces of alcohol daily during pregnancy.

    a. 2
    b. 3
    c. 4
    d. 5

16. Producing 25 ounces of milk a day costs a woman almost _____ calories per day.

    a. 300
    b. 500
    c. 650
    d. 1000

17. The birthweight of an infant triples by the age of:

    a. 4 months.
    b. 6 months.
    c. 9 months.
    d. 1 year.

18. Reduced-fat milk is not recommended for infants and children before the age of:

    a. 6 months.
    b. 2 years.
    c. 3 years.
    d. 4 years.

19. Water should be offered regularly to both breastfed and bottle-fed infants.

    a. true
    b. false

20. WIC participation during pregnancy can effectively reduce infant mortality, low birthweight, and maternal and newborn medical costs.

    a. true
    b. false

21. All of the following are appropriate physical activities for the pregnant woman **except:**

    a. water aerobics.
    b. light strength training.
    c. climbing stairs.
    d. volleyball.

22. The greatest risk for pregnant teenagers is:

    a. having low-birthweight infants.
    b. the death of the infant within the first year.
    c. experiencing a miscarriage.
    d. having a premature baby.

23. Breastfeeding is inadvisable if the mother:

    a. is drinking alcohol.
    b. has a cold.
    c. is on drugs.
    d. a and b
    e. a and c

24. Breastfeeding protects against common illnesses of infancy such as middle ear infection and respiratory illness.

    a. true
    b. false

Application Level Items:

25. Lola experienced malnutrition late in her pregnancy. Which of the following organs of the infant will most likely be affected?

    a. heart
    b. lungs
    c. brain
    d. a and c
    e. b and c

26. Cindy is a 25-year-old pregnant female who recently experienced a sudden weight gain. In addition, she has blurred vision, is dizzy, and has frequent headaches. Cindy is most likely experiencing symptoms of:

    a. malnutrition.
    b. gestational diabetes.
    c. preeclampsia.
    d. a normal pregnancy.

27. A nonpregnant woman requires 2200 calories to maintain her desirable body weight. How many calories would she need if she were in her third trimester of pregnancy?

    a. 2100
    b. 2300
    c. 2500
    d. 2650

28. Joan is a strict vegetarian who is now pregnant. Which of the following would you recommend as appropriate sources of protein?

    a. legumes
    b. tofu
    c. protein supplements
    d. a and b
    e. b and c

29. Jane weighs 160 pounds (prepregnancy) and is considered to be obese. Approximately how much should Jane weigh at the end of her pregnancy?

    a. 160
    b. 175
    c. 180
    d. 190

30. Which of the following indicates an infant who is ready for solid foods?

    a. Jeff, who is 4 months old
    b. Claudia, who can sit with support and control head movements
    c. Ben, who swallows using the back of his tongue
    d. Sarah, who turns her head toward any object that brushes her cheek

Answers

Summing Up

(1) nourish; (2) weight; (3) birthweight; (4) health; (5) placenta; (6) sustenance; (7) critical; (8) heart; (9) digestive; (10) 40; (11) birth; (12) 340; (13) 25; (14) carbohydrate; (15) folate; (16) phosphorus; (17) iron; (18) three; (19) pound; (20) 500; (21) 13; (22) alcohol; (23) water; (24) breast; (25) vitamin D; (26) six; (27) six; (28) readiness; (29) development; (30) weight

Chapter Glossary

Matching Exercise: (1) g; (2) c; (3) l; (4) d; (5) k; (6) a; (7) i; (8) f; (9) j; (10) b; (11) h; (12) e; (13) o; (14) n; (15) m; (16) q; (17) p
Crossword Puzzle: (1) fetal alcohol syndrome; (2) preeclampsia; (3) spina bifida; (4) gestational diabetes; (5) alpha-lactalbumin; (6) lactation; (7) cesarean section; (8) ARND; (9) anencephaly; (10) lactoferrin; (11) neural tube defects; (12) milk anemia; (13) edema; (14) colostrum

Exercises

Chapter Study Questions:

1. A strong correlation exists between prepregnancy weight, weight gain during pregnancy, and infant birthweight. Infant birthweight is the most potent single indication of the infant's future health status. Low-birthweight babies have a greater chance of dying early in life and of having illnesses.

2. No; many women in the United States exceed the recommended protein intake for pregnancy even when they are not pregnant and excessive protein intake may have adverse effects.

3. Menstruation ceases and the absorption of iron increases up to threefold. However, because iron stores dwindle, blood is lost during birth, and few women enter pregnancy with adequate stores to meet pregnancy demands, a supplement is recommended during the second and third trimesters.

4. In general, the effect of nutritional deprivation is to reduce the quantity, not the quality, of milk. However, the levels of some vitamins in human milk can be affected by excessive or deficient intakes of the mother.

5. A variety of nutritious foods should be introduced in an inviting way; the baby should not be forced to finish a bottle or jar of baby food; concentrated sweets and empty-calorie foods should be limited; babies should not be taught to seek food as a reward and they should not be comforted or punished with food; physical activity should be encouraged.

6. Because they contain high concentrations of mercury which is an environmental contaminant. Infants exposed to these contaminants show signs of impaired cognitive development. Mercury readily moves across the placenta and affects the developing fetal nervous system.

Short Answer Questions:

1. (a) poor nutrition; (b) heredity; (c) disease conditions; (d) smoking; (e) drug use

2. (a) defects of the nervous system of the embryo; (b) poor dental health of children whose mothers were malnourished during pregnancy; (c) adult's and adolescent's vulnerability to infections and possibly higher risks of diabetes, hypertension, stroke and heart disease

3. (a) the great increase in the number of the mother's red blood cells; (b) new cells being laid down at a tremendous pace as the fetus grows and develops

4. (a) calcium; (b) phosphorus; (c) magnesium

5. (a) folate; (b) iron; (c) calcium

6. (a) placenta; (b) uterus; (c) blood; (d) milk-producing glands; (e) fetus itself

7. (a) a few sips of a carbonated drink; (b) a few nibbles of soda crackers or other salty snack foods

8. (a) diarrhea; (b) vomiting

9. (a) the baby's nutrient needs; (b) the baby's physical readiness to handle different forms of foods; (c) need to detect and control allergic reactions

10. (a) supplying nutritious foods; (b) avoiding sweets; (c) discouraging the association of food with reward or comfort; (d) discouraging the use of a bottle as a pacifier

Problem-Solving:

1. (a) caloric needs: 2000 calories + 340 calories = 2340 calories; (b) recommended carbohydrate intake = 135-175 grams; (c) recommended protein intake = .8 g/kg body weight + 25 additional grams of protein = .8 × 59 = 47 + 25 = 72 grams

2. c
3. a
4. e
5. b
6. d
7. b
8. e
9. d
10. a
11. c
12. f

Controversy Questions:

1. They are female, have a family history of type 2 diabetes, are of non-European descent, have mothers who had diabetes while pregnant, have metabolic syndrome, have a low family income, and are more sedentary than their peers.

2. Type 2 diabetes in children is associated with obesity, sedentary behavior, and a family history of diabetes.

3. Physically active children have a better lipid profile and lower blood pressure than sedentary children.

4. A diet that provides 25-35% of total energy as fat, and restriction of saturated fat and trans fat are appropriate.

5. Participation in regular aerobic activity and losing weight or maintaining weight as children grow taller.

Study Aids

1. (a) infant formula; (b) cereal; (c) meats; (d) vegetables; (e) meats; (f) eggs

Sample Test Items

1. d (p. 482)	11. b (p. 505-506)	21. d (p. 493)
2. b (p. 482-483)	12. d (p. 507)	22. b (p. 493)
3. c (p. 482)	13. c (p. 506)	23. e (p. 500-501)
4. a (p. 487)	14. e (p. 493)	24. a (p. 506)
5. a (p. 489)	15. a (p. 499)	25. b (p. 484)
6. c (p. 490)	16. b (p. 500)	26. c (p. 491, 499)
7. c (p. 491, 499)	17. d (p. 503)	27. d (p. 485)
8. b (p. 496)	18. b (p. 507)	28. d (p. 487)
9. e (p. 494)	19. b (p. 503)	29. b (p. 491)
10. c (p. 500)	20. a (p. 490-491)	30. b (p. 509)

Chapter 14 - Child, Teen, and Older Adult

Chapter Objectives

After completing this chapter, you should be able to:

1. Describe the nutrient needs of young children and appropriate feeding practices including such issues as choking, portion sizes, and snacking.

2. Discuss nutrition-related concerns of children including the link between diet and behavior, the problem of lead, and the impact of television on nutrition.

3. Distinguish between food allergies, intolerances, and aversions.

4. Discuss the special nutrient needs and concerns of teenagers including the effect of diet on PMS and acne.

5. Discuss special nutritional needs of older adults and the suspected connections between diet and disease.

6. Discuss common nutrient-drug interactions (Controversy 14).

Key Concepts

✓ Children's nutrient needs reflect their stage of growth. For a healthy child, use the DRI recommended intakes, the *Dietary Guidelines for Americans 2005*, and the *Food Guide Pyramid for Young Children* as guides to establish food patterns that provide adequate nourishment for growth while defending against obesity and chronic disease.

✓ Healthy eating habits and positive relationships with food are learned in childhood. Parents teach children best by example. Choking can often be avoided by supervision during meals and avoiding hazardous foods.

✓ The detrimental effects of nutrient deficiencies in children in developed nations can be subtle. Iron deficiency is the most widespread nutrition problem of children and causes abnormalities in both physical health and behavior. Iron toxicity is a major form of poisoning in children.

✓ Lead poisoning has declined dramatically over the past two decades, but when it occurs it can inflict severe, irreparable damage on growing children. Higher awareness of the remaining sources of lead poisoning can help to reduce the present rate of occurrence.

✓ Food allergies can cause serious illness. Diagnosis is based on presence of antibodies, and tests are imperative to determine whether allergy exists. Food aversions can be related to food allergies or to adverse reactions to food.

✓ Hyperactivity, properly named attention-deficit/hyperactivity disorder (ADHD), is not caused by food allergies; temporary "hyper" behavior may reflect excess caffeine consumption or inconsistent care. A wise parent will limit children's caffeine intakes and meet their needs for structure. Poverty may cause behavior problems.

✓ The nation's children are growing fatter and face advancing risks of diseases. Childhood obesity demands careful family-centered management. Television viewing can contribute to obesity through lack of exercise and promoting overconsumption of calorie-dense snacks. The sugary snacks advertised on children's TV programs also contribute to dental caries.

- ✓ Breakfast is critical to school performance. Not all children start the day with an adequate breakfast, but school breakfast programs help to fill the need.

- ✓ School lunches are designed to provide at least a third of the nutrients needed daily by growing children and to stay within limits set by the *Dietary Guidelines for Americans*. Soda and snack vending machines, fast-food and snack bars, and school stores tempt students with foods high in fats and sugars. Fruit juice is a healthy food but may cause dental problems if used to excess.

- ✓ The need for iron increases during adolescence for males and females. Women's iron needs remain higher as long as they continue to menstruate.

- ✓ Sufficient calcium intake is crucial during adolescence to support normal bone growth and density. When teens choose soft drinks and abandon milk, they increase their chances of bone disease later on in life.

- ✓ The adolescent growth spurt increases the need for energy and nutrients. The normal gain of body fat during adolescence may be mistaken for obesity, particularly in girls. Some self-prescribed diets are detrimental to health and growth.

- ✓ Although no foods have been proved to aggravate acne, stress can worsen it. Supplements are useless against acne, but sunlight, proven medications, and relief from stress can help.

- ✓ With planning, the gatekeeper can encourage teens to meet nutrient requirements by providing nutritious snacks.

- ✓ Life expectancy for U.S. adults increased in the twentieth century. Life choices can greatly affect how long a person lives and the quality of life in the later years. No diet or supplement can extend life.

- ✓ Energy needs decrease with age, but exercise burns off excess fuel, maintains lean tissue, and brings health benefits.

- ✓ Protein needs remain about the same through adult life, but choosing low-fat fiber-rich protein foods may help control other health problems.

- ✓ Generous carbohydrate intakes are recommended for older adults. Including fiber in the diet is important to avoid constipation.

- ✓ A diet high in fruits and vegetables and low in fats of meats and dairy products may improve some symptoms of arthritis. Omega-3 fatty acids may also have a positive effect.

- ✓ Vitamin A absorption increases with aging. Older people suffer more from deficiencies of vitamin D and vitamin B_{12} than young people do. Cataracts and macular degeneration often occur among those with low fruit and vegetable intakes.

- ✓ Aging alters vitamin and mineral needs. Some needs rise while others decline.

- ✓ Lifestyle factors can make a difference in aging. In rats and other species, food energy deprivation may lengthen the lives of individuals. Claims for life extension through antioxidants or other supplements are common hoaxes.

- ✓ Alzheimer's disease causes some degree of brain deterioration in many people past age 65. Current treatment helps only marginally; dietary aluminum is probably unrelated. The importance of nutrition care increases as the disease progresses.

✓ Food choices of the elderly are affected by aging, altered health status, and changed life circumstances. Assistance programs can help by providing nutritious meals, offering opportunities for social interactions, and easing financial problems.

Summing Up

Nutrient needs change throughout life and vary depending on the rate of growth, (1)_____, activities, and many other factors. An infant's appetite decreases near the (2)_____ birthday, in line with the great reduction in growth rate. A one-year-old child needs about (3)_____ calories a day; a six year old needs only (4)_____ calories more. To prevent iron deficiency, children's foods must deliver approximately (5)_____ milligrams of iron per day.

"Hyper" behavior in children may reflect excess (6)_____ consumption or inconsistent care. Excess caffeine may be consumed by children in the form of tempting (7)_____ and chocolate. Research has found that obesity in children may be related to (8)_____ viewing, which affects children's nutritional health adversely in several ways.

With the onset of (9)_____, needs for all nutrients become greater than at any other time of life except during pregnancy and lactation. In females, (10)_____ becomes a larger percentage of the total body weight. On the average, about a fourth of a teenager's total daily energy intake comes from (11)_____. One of the many changes girls face as they become women is the onset of (12)_____. The hormones that regulate the menstrual cycle also affect metabolic rate, glucose tolerance, (13)_____, food intake, mood, and (14)_____.

Teenagers often fail to obtain enough (15)_____ because the need for this nutrient increases during the growth spurt regardless of the age of the adolescent. Another problem nutrient for teenagers is (16)_____, which may fall short unless they snack on dairy products. Although no foods have been proven to aggravate acne, (17)_____ can worsen it.

In the U.S., the life expectancy at birth is (18)_____ years for white women and 75 years for white men. Nutrient needs become more individual with age, depending on genetics and medical history, but (19)_____ needs often decrease with advancing age. A steady supply of (20)_____ is essential for older adults for optimal brain functioning. A diet high in fruits and vegetables and low in (21)_____ of meats and dairy products may improve some symptoms of arthritis.

(22)_____ is a major risk for older adults, who may not notice or pay attention to their thirst. With age, (23)_____ takes on extra importance for its role against constipation. Among the vitamins, vitamin (24)_____ stands alone in that absorption appears to increase with aging. Older adults also face a greater risk of vitamin (25)_____ deficiency because its synthesis declines fourfold as people age. Cataracts are more likely to occur in people with low fruit and green (26)_____ intakes. Iron deficiency is less common in older adults than in younger people, but (27)_____ deficiencies are common in older people and may lead to decreased appetite and a diminished sense of (28)_____. Preventing (29)_____ is an important concern for the older person suffering with Alzheimer's disease. Two factors that make older people vulnerable to malnutrition include use of multiple (30)_____ and abuse of alcohol.

Chapter Glossary

Matching Exercise:

_____	1.	premenstrual syndrome
_____	2.	growth spurt
_____	3.	anaphylactic shock
_____	4.	food aversion
_____	5.	allergy
_____	6.	longevity
_____	7.	senile dementia
_____	8.	life span
_____	9.	hyperactivity
_____	10.	life expectancy
_____	11.	acne
_____	12.	histamine
_____	13.	antigen
_____	14.	cataracts
_____	15.	food intolerance
_____	16.	antibodies
_____	17.	arthritis
_____	18.	learning disabilities
_____	19.	epiphyseal plate
_____	20.	epinephrine

a. a syndrome characterized by inattention, impulsiveness, and excess motor activity

b. a life-threatening whole-body allergic reaction to an offending substance

c. an adverse reaction to a food or food additive not involving the immune response

d. an immune reaction to a foreign substance, such as a component of food

e. a substance foreign to the body that elicits the formation of antibodies or an inflammation reaction from immune system cells

f. a cluster of symptoms that some women experience prior to and during menstruation

g. the loss of brain function beyond the normal loss of physical adeptness and memory that occurs with aging

h. chronic inflammation of the skin's follicles and oil-producing glands, which leads to an accumulation of oils inside the ducts that surround hairs

i. large protein molecules that are produced in response to the presence of antigens and then help to inactivate the antigens

j. a hormone of the adrenal gland that counteracts anaphylactic shock by opening the airways and maintaining heartbeat and blood pressure

k. a substance that participates in causing inflammation; produced by cells of the immune system as part of a local immune reaction to an antigen

l. an intense dislike of a food, biological or psychological in nature, resulting from an illness or other negative experience associated with that food

m. the average number of years lived by people in a given society

n. clouding of the lens of the eye that can lead to blindness

o. long duration of life

p. a usually painful inflammation of the joints caused by many conditions, including infections, metabolic disturbances, or injury

q. the maximum number of years of life attainable by a member of a species

r. a thick cartilage-like layer that forms new cells that are eventually calcified, lengthening the bone

s. a group of conditions resulting in an altered ability to learn basic cognitive skills such as reading, writing, and mathematics

t. the marked rapid gain in physical size usually evident around the onset of adolescence

Exercises

Answer these chapter study questions:

1. Identify four ways in which television affects children's nutritional health adversely.

2. Compare the characteristics of children who eat no breakfast with those of their well-fed peers.

3. What recommendations would you make to an individual who experiences PMS?

4. Why should older adults attend to fat intakes?

5. Why do energy needs decrease with age?

6. How would you prevent a child from choking?

Complete these short answer questions:

1. School lunches must include specified servings of:

   a.
   b.
   c.
   d.
   e.

2. Allergies may have these one or two components:

   a.
   b.

3. The life threatening food allergy reaction of anaphylactic shock is most often caused by these foods:

   a.   e.
   b.   f.
   c.   g.
   d.   h.

4. Normal, everyday causes of "hyper" behavior in children include:

   a.
   b.
   c.
   d.
   e.
   f.
   g.

5. Hormones that regulate the menstrual cycle also alter:

   a.
   b.
   c.
   d.
   e.
   f.

6. Deficiencies of these vitamins and minerals enhance lead absorption:

   a.
   b.
   c.
   d.
   e.

7. Minerals frequently lacking in older people's diets include:

   a.
   b.
   c.

8. Older people face a greater risk of vitamin D deficiency than younger people do because:

   a.
   b.
   c.

9. Six factors that affect physiological age include:

   a.
   b.
   c.
   d.
   e.
   f.

10. The defining symptom of Alzheimer's disease is impairment of memory and reasoning powers, but it may also be accompanied by:

   a.   f.
   b.   g.
   c.   h.
   d.   i.
   e.

Solve these problems:

1. Below is a sample day's menu for a 4- to 6-year-old child. Compare the menu with the *Food Guide Pyramid for Young Children* and identify foods and potential nutrients which would be missing from the child's diet.

   ### Menu

   Breakfast: Cheese toast with 1 slice of whole-wheat bread and 1 ounce of cheddar cheese; ¾ cup orange juice; ½ cup whole milk

   Lunch: Peanut butter and jelly sandwich with 4 tbsp. peanut butter and 2 tbsp. jelly on 2 slices of whole-wheat bread; 1 apple; 1 cup whole milk

   Dinner: 2 ounces of steak; ½ cup mashed potato with 1 tbsp. butter; 1 dinner roll; 1 cup strawberry yogurt

   Snack: 3 chocolate chip cookies

2. Identify whether the following foods have high or low caries potential by placing an X in the appropriate space.

		Low Caries Potential	High Caries Potential
a.	plain yogurt	_____	_____
b.	sugared cereals	_____	_____
c.	bagels	_____	_____
d.	dried fruit	_____	_____
e.	popcorn	_____	_____
f.	glazed carrots	_____	_____
g.	sugarless gum	_____	_____
h.	legumes	_____	_____
i.	fresh fruit	_____	_____
j.	chocolate milk	_____	_____

3. For each nutrient listed below, indicate whether nutritional needs for older adults are higher, lower, or the same as that for younger adults.

		Higher	Lower	Same
a.	energy	_____	_____	_____
b.	fiber	_____	_____	_____
c.	vitamin A	_____	_____	_____
d.	vitamin D	_____	_____	_____
e.	protein	_____	_____	_____

Answer these controversy questions:

1. Describe characteristics of individuals who have the greatest potential for undesirable nutrient-drug interactions.

2. How can chronic use of laxatives, such as mineral oil, lead to malnutrition?

3. Why is iron-deficiency anemia likely to occur in people who take large quantities of aspirin?

4. Why are women older than 35 years who take oral contraceptives at increased risk of cardiovascular disease?

5. How is moderation in caffeine consumption defined by health care authorities?

Study Aids

Study Figure 14-3 on page 523 of your textbook to identify the food groups, number of servings recommended, and food serving sizes in the *Food Guide Pyramid for Young Children*. Study Table 14-7 on page 544 in your textbook to identify nutrient concerns that accompany aging.

Sample Test Items

Comprehension Level Items:

1. The most common nutrient deficiency among children and adolescents in the United States is:

   a. calcium deficiency.
   b. iron deficiency.
   c. vitamin D deficiency.
   d. vitamin A deficiency.

2. Television's influence on nutrition is evident in:

   a. childhood obesity.
   b. dental health of children.
   c. childhood eating disorders.
   d. a and b
   e. b and c

3. Which of the following foods most often causes allergic reactions?

   a. cheese
   b. poultry
   c. peanuts
   d. rye

4. Which of the following is true of children who eat no breakfast compared to those who do?

   a. They perform poorly in tasks of concentration.
   b. Their attention spans are longer.
   c. They achieve higher test scores than their well-fed peers.
   d. They have the same rates of obesity as their well-fed peers.

5. Needs for all nutrients are greater during _____ than at any other time of life except during pregnancy and lactation.

   a. infancy
   b. childhood
   c. adolescence
   d. adulthood

6. On the average, approximately one-fourth of teenagers' total daily energy intake comes from:

   a. snacks.
   b. the school lunch program.
   c. meals.
   d. fast food meals.

7. To get children to accept vegetables you would:

   a. serve vegetables which are easy to eat.
   b. serve vegetables that are highly flavored.
   c. serve brightly colored vegetables.
   d. a and b
   e. a and c

8. Prevention of lead toxicity in children rests primarily on:

   a. reduced exposure.
   b. adequate intake of calcium.
   c. increased awareness of mothers.
   d. adequate intake of iron.

9. Which of the following foods have been shown not to worsen acne?

   a. chocolate
   b. fatty acids
   c. sugar
   d. a and b
   e. a and c

10. Which of the following would you recommend to a woman who experiences PMS?

   a. a diuretic
   b. vitamin E supplement
   c. a caffeine-free lifestyle
   d. a fiber-restricted diet

11. Which of the following protein sources would you recommend to an older adult with constipation?

    a. beef
    b. eggs
    c. legumes
    d. milk

12. In the United States, the life expectancy for black men is:

    a. 68 years.
    b. 79 years.
    c. 80 years.
    d. 115 years.

13. A diet low in _____ may improve some symptoms of arthritis.

    a. protein
    b. sodium
    c. sugar
    d. saturated fat

14. Energy needs often decrease with advancing age.

    a. true
    b. false

15. As a person grows older:

    a. dehydration becomes a major risk.
    b. fiber takes on extra importance.
    c. protein needs increase.
    d. a and b
    e. b and c

16. Which of the following statements is true concerning the dietary habits of older adults?

    a. They have cut down on saturated fats.
    b. They are eating fewer vegetables.
    c. They are consuming more milk.
    d. They are eating fewer whole-grain breads.

17. There is an apparent increase in the absorption of vitamin _____ with aging.

    a. C
    b. D
    c. A
    d. E

18. Which of the following is provided by the Senior Nutrition Program?

    a. nutritious meals
    b. transportation
    c. assistance with housekeeping
    d. a and b
    e. b and c

19. Which of the following puts older adults at greatest risk of developing cataracts according to the latest scientific studies?

    a. excess food energy intake
    b. diets low in fruits and green vegetables
    c. excess intakes of milk and sugar
    d. deficiencies of riboflavin and zinc

20. Internal appetite regulation in children of normal weight guarantees that their overall energy intakes will be right for each stage of growth.

    a. true
    b. false

21. To determine the appropriate fiber intake for children you would add the number _____ to their age.

    a. 2
    b. 3
    c. 4
    d. 5

22. Which of the following are ways to prevent iron deficiency in children?

    a. Provide 10 milligrams of iron per day from foods.
    b. Encourage the consumption of four servings of milk per day.
    c. Serve iron-rich foods at meals and snacks.
    d. a and b
    e. a and c

23. Research has shown a positive correlation between childhood obesity and fruit juice consumption.

    a. true
    b. false

24. Teenagers often choose _____ as their primary beverage.

    a. water
    b. milk
    c. soft drinks
    d. fruit juice

Application Level Items:

25. Which of the following would the best snack choice for an active, normal weight child?

    a. flavored gelatin
    b. candy
    c. cola
    d. oatmeal cookies

26. Which of the following vegetables would be the most acceptable to a child?

    a. slightly overcooked green beans
    b. mixed vegetables
    c. broccoli
    d. bright green peas

27. Betsy is a two year old who has been on a food jag for five days. What is the best way to respond to Betsy's food jag?

    a. no response, since attention is a valuable reward
    b. serve tiny portions of many foods
    c. serve favored food items
    d. distract the child with friends at meals

28. To control dental caries, these foods should be used infrequently:

    a. cheese, apples, and hard rolls.
    b. toast, chicken, and green beans.
    c. chocolate milk, raisins, and candied sweet potatoes.
    d. plain yogurt, eggs, and pizza.

29. Which of the following snack foods would be most appropriate for a teenager trying to consume enough vitamin A?

    a. hard boiled eggs, bran muffin, and crackers
    b. carrot sticks, tomato juice, and dried apricots
    c. luncheon meat, whole-wheat bread, and celery sticks
    d. Swiss cheese, crackers, and cantaloupe

30. Bessie is an 85-year-old female who resides in a nursing home. She has a good appetite and consumes three cups of milk a day. However, Bessie is at risk for developing vitamin D deficiency because:

    a. she does not have adequate exposure to sunlight.
    b. she does not consume an adequate amount of milk.
    c. as she ages, her vitamin D synthesis declines.
    d. a and b
    e. a and c

Answers

Summing Up

(1) gender; (2) first; (3) 800; (4) 800; (5) 7-10; (6) caffeine; (7) colas; (8) television; (9) adolescence; (10) fat; (11) snacks; (12) menstruation; (13) appetite; (14) behavior; (15) iron; (16) calcium; (17) stress; (18) 80; (19) energy; (20) carbohydrate; (21) fats; (22) dehydration; (23) fiber; (24) A; (25) D; (26) vegetable; (27) zinc; (28) taste; (29) weight loss; (30) medications

Chapter Glossary

Matching Exercise: (1) f; (2) t; (3) b; (4) l; (5) d; (6) o; (7) g; (8) q; (9) a; (10) m; (11) h; (12) k; (13) e; (14) n; (15) c; (16) i; (17) p; (18) s; (19) r; (20) j

Exercises

Chapter Study Questions:

1. Television viewing requires no energy and seems to reduce the metabolic rate to levels below that of rest and it contributes to physical inactivity by consuming time that could be spent in more vigorous activities. Watching television also correlates with between-meal snacking and with eating the calorically dense and fatty foods most heavily advertised on children's programs. Finally, children who watch more television or watch it during meals are least likely to eat fruits and vegetables.

2. Children who eat no breakfast are more likely to be overweight and to perform poorly in tasks of concentration, their attention spans are shorter, they achieve lower test scores, and they are tardy or absent more often than their well-fed peers.

3. The individual should try to examine her total lifestyle, including diet and stress, get adequate sleep, engage in physical activity, and try to moderate her intake of caffeine.

4. Foods low in saturated and trans fats are often richest in vitamins, minerals and phytochemicals and may help retard the development of atherosclerosis, obesity, and other diseases. A diet low in the fats of meats and dairy products may also improve some symptoms of arthritis. However, enough of the essential fatty acids supports good health throughout life.

5. The number of active cells in each organ decreases, which brings about a reduction in the body's metabolic rate. In addition, older people are not usually as physically active as younger people and their lean tissue diminishes.

6. Have the child sit while eating and eliminate round foods, such as grapes and hard candies, or other dangerous foods, such as popcorn, tough meat, and peanut butter eaten by the spoonful.

Short Answer Questions:

1. (a) milk; (b) protein-rich foods; (c) vegetables; (d) fruits; (e) breads or other grains

2. (a) antibodies; (b) symptoms

3. (a) milk; (b) eggs; (c) peanuts; (d) tree nuts; (e) wheat; (f) soybeans; (g) fish; (h) shellfish

4. (a) desire for attention; (b) lack of sleep; (c) overstimulation; (d) too much television; (e) lack of exercise; (f) chronic hunger; (g) too much caffeine from colas or chocolate

5. (a) the metabolic rate; (b) glucose tolerance; (c) appetite; (d) food intake; (e) mood; (f) behavior

6. (a) iron; (b) calcium; (c) zinc; (d) vitamin D; (e) vitamin C

7. (a) iron; (b) zinc; (c) calcium

8. (a) many drink little or no vitamin D-fortified milk; (b) many get little or no exposure to sunlight; (c) as people age, vitamin D synthesis declines

9. (a) abstinence from, or moderation in, alcohol use; (b) regular meals; (c) weight control; (d) adequate sleep; (e) abstinence from smoking; (f) regular physical activity

10. (a) anxiety; (b) loss of ability to communicate; (c) loss of physical capabilities; (d) delusions; (e) depression; (f) inappropriate behavior; (g) irritability; (h) sleep disturbance; (i) eventual loss of life

Problem-Solving:

1. About 3 servings were consumed in the milk and milk product groups, which exceeds the two recommended servings. Two servings of meat foods were consumed, making this group adequate. The bread and cereal food group was not adequately represented, with a total of four servings versus the recommended six. Two servings of fruits were consumed, which was adequate, and only one vegetable was consumed, versus the three recommended. As a consequence, vitamins A and C, folate, carbohydrate, and fiber intake could be low for the day.

2.

		Low Caries Potential	*High Caries Potential*
a.	plain yogurt	x	
b.	sugared cereals		x
c.	bagels	x	
d.	dried fruit		x
e.	popcorn	x	
f.	glazed carrots		x
g.	sugarless gum	x	
h.	legumes	x	
i.	fresh fruit	x	
j.	chocolate milk		x

3.

		Higher	*Lower*	*Same*
a.	energy		x	
b.	fiber			x
c.	vitamin A		x	
d.	vitamin D	x		
e.	protein			x

Controversy Questions:

1. Those who take drugs for long times; those who take 2 or more drugs at the same time; and those who are poorly nourished to begin with or not eating well.

2. Laxatives can carry nutrients through the intestines so rapidly that many vitamins have no time to be absorbed. Also, mineral oil can rob a person of fat-soluble vitamins, and calcium can also be excreted along with the oil, leading to adult bone loss.

3. Aspirin speeds up blood loss from the stomach by as much as ten times, leading to iron-deficiency anemia.

4. Because oral contraceptives alter blood lipids by raising total cholesterol and triglyceride concentrations and by lowering HDL.

5. Moderation is defined as the equivalent of 2 small cups of coffee per day.

Sample Test Items

1. b (p. 522, 526)
2. d (p. 530)
3. c (p. 528)
4. a (p. 531)
5. c (p. 534)
6. a (p. 536)
7. e (p. 524)
8. a (p. 526-527)
9. e (p. 536)
10. c (p. 537)
11. c (p. 540-541)
12. a (p. 538)
13. d (p. 541)
14. a (p. 538)
15. d (p. 541, 543)
16. a (p. 547)
17. c (p. 542)
18. d (p. 548)
19. b (p. 542)
20. a (p. 521)
21. d (p. 522)
22. e (p. 522)
23. b (p. 532)
24. c (p. 534)
25. d (p. 525)
26. d (p. 524)
27. a (p. 524)
28. c (p. 531)
29. b (p. 536)
30. e (p. 542)

Chapter 15 - Hunger and the Global Environment

Chapter Objectives

After completing this chapter, you should be able to:

1. Discuss some of forces that threaten world food production and distribution in the next decade.

2. Distinguish between the following terms: hunger, food insecurity, food poverty, and food shortage.

3. List some of the U.S. government programs designed to relieve poverty and hunger and discuss their effectiveness.

4. Compare the causes of hunger in the United States and in developing countries and explain the roles played by food shortages, chronic hunger, and the diminishing world food supply.

5. Explain how relieving poverty and overpopulation will help alleviate environmental degradation and hunger.

6. List some of the environmentally conscious food shopping and cooking activities consumers can choose to minimize the impacts on the environment.

7. Discuss some of the environmental and social costs of agriculture and food production and indicate possible solutions that would enable both agribusiness and the environment to survive (Controversy 15).

Key Concepts

✓ The world's chronically hungry people suffer the effects of undernutrition, and many in the United States live with food insecurity. Many forces combine to threaten the world's future food supply and its distribution.

✓ Chronic hunger causes many deaths worldwide, especially among children. Intermittent hunger is frequently seen in U.S. children. The immediate cause of hunger is poverty.

✓ Poverty and hunger coexist with affluence and bounty in the United States; working people as well as the unemployed experience poverty and hunger. Government programs to relieve poverty and hunger are tremendously helpful, if not fully successful.

✓ Natural causes such as drought, flood, and pests and political and social causes such as armed conflicts and overpopulation all contribute to hunger and poverty in developing countries. To meet future demands for food, technology must continue to improve food production, food must be fairly distributed, and birthrates need to decline. The world's women and girls are major allies in the effort to fight hunger.

✓ Environmental degradation caused by the impacts of growing numbers of people is threatening the world's future ability to feed all of its citizens. Improvements in agriculture can no longer keep up with people's growing numbers. Human population growth is an urgent concern. Controlling population growth requires improving people's economic status and providing them with health care, education, and family planning.

✓ Government, business, educators, and all individuals have many opportunities to promote sustainability worldwide and wise resource use at home.

Summing Up

In the United States, over (1)_____ million children do not know where their next meal is coming from or when it will come. (2)_____ is uncertain or limited access to foods of sufficient quality or quantity to sustain a healthy and active life. The primary cause of hunger in developed nations is (3)_____. In the U.S. food poverty affects not only the chronic poor but also the so-called (4)_____ poor.

Hunger and poverty interact with a third force, which is (5)_____ degradation. One element of this degradation is (6)_____ which is affecting agriculture in every nation. The only way to enable the world's output to keep pace with people's growing numbers is (7)_____ control.

Population growth, (8)_____ and environmental degradation combine to worsen each other. Breaking the cycle requires improving the (9)_____ status of the people and providing them with health care, (10)_____, and family planning. Both rich and poor nations play key roles in solving the world's environmental, poverty, and (11)_____ problems. Poor nations need to make (12)_____ technology and information more widely available, (13)_____ their citizens and adopt (14)_____ development practices that slow the destruction of their forests, waterways and soil. Rich nations need to stem their wasteful and polluting uses of resources and (15)_____, which are contributing to global environmental degradation.

Many nations now agree that improvement of all nation's (16)_____ is a prerequisite to meeting the world's other urgent needs. At a summit of over 100 nations, many agreed to a set of principles of (17)_____ development. The conference defined sustainable development as development that would equitably meet both the economic and (18)_____ needs of present and future generations.

Consumers can choose to minimize negative impacts on the (19)_____ through their daily choices in many ways. Small decisions each day, such as making fewer (20)_____ trips, choosing foods low on the (21)_____, and buying locally grown foods, can add up to large impacts on the environment. In addition, consumers can buy bulk items with minimal (22)_____, use (23)_____ cooking methods, and purchase the most efficient large (24)_____ possible. Products bearing the U.S. government's (25)_____ logo rank highest for energy efficiency.

Chapter Glossary

Matching Exercise:

_____ 1. food poverty

_____ 2. famine

_____ 3. food insecurity

_____ 4. world food supply

_____ 5. food recovery

_____ 6. hunger

_____ 7. sustainable

_____ 8. oral rehydration therapy

_____ 9. carrying capacity

_____ 10. food shortage

_____ 11. food banks

_____ 12. food pantries

a. lack or shotage of basic foods needed to provide energy and nutrients that support health

b. oral fluid replacement for children with severe diarrhea caused by infectious disease

c. extreme scarcity of food in an area that causes starvation and death in a large portion of the population

d. community food collection programs that provide groceries to be prepared and eaten at home

e. hunger occurring when an area of the world lacks enough total food to feed its people

f. the quantity of food, including stores from previous harvests, available to the world's people at a given time

g. able to continue indefinitely; the term refers to the use of resources at such a rate that the earth can keep on replacing them

h. hunger occurring when enough food exists in an area but some of the people cannot obtain it because they lack money, they are being deprived for political reasons, they live in a country at war, or because of other problems such as lack of transportation

i. the total number of living organisms that a given environment can support without deteriorating in quality

j. limited or uncertain access to foods of sufficient quality or quantity to sustain a healthy and active life

k. collecting wholesome surplus food for distribution to low-income people who are hungry

l. a facility that collects and distributes food donations to authorized organizations feeding the hungry

Exercises

Answer these chapter study questions:

1. What is meant by the term *sustainable* and what is an example?

2. Why is it environmentally beneficial to eat low on the food chain?

3. Why are fast cooking methods recommended as one step consumers should take to minimize negative impacts on the environment? What are some examples?

4. Describe what is meant by the statement that hunger, poverty, and environmental degradation all interact with each other.

5. Why do people living in poverty bear more children?

6. Why are women being targeted as direct recipients of food relief?

Complete these short answer questions:

1. Some of the forces threatening world food production and distribution in the next decade include:

a.   f.
b.   g.
c.   h.
d.   i.
e.   j.

2. The primary cause of hunger in developed nations is _____.

3. Natural causes of famine include:

   a.
   b.
   c.

4. Fast cooking methods that use less energy than conventional stovetop cooking methods include:

   a.
   b.
   c.

5. To minimize the energy a refrigerator uses you would:

   a.
   b.
   c.

6. To solve the world's environmental, poverty, and hunger problems, the poor nations need to:

   a.
   b.
   c.
   d.

7. The death rate falls when people attain better access to:

   a.
   b.
   c.

8. Food stamp coupons cannot be used to purchase:

   a.
   b.
   c.
   d.

9. To solve the world's environmental, poverty, and hunger problems, the rich nations need to

   _____.

10. In the U.S., the chronic poor include:

    a.
    b.
    c.
    d.

Solve these problems:

Place an X next to each of the behaviors below which are environmentally conscious.

_____ car pool to the grocery store
_____ eat high on the food chain
_____ buy foods which are locally grown
_____ buy canned beef products
_____ select fish such as tuna, swordfish, and shark
_____ buy items in bulk
_____ use plastic bags instead of paper ones
_____ use the microwave for cooking

Answer these controversy questions:

1. What is agribusiness?

2. How does massive fossil fuel use threaten our planet?

3. What is the problem with raising animals in concentrated areas such as cattle feed lots?

4. What is meant by *integrated pest management*?

5.   How can biotechnology lead to economic, environmental, and agricultural benefits?

Sample Test Items

Comprehension Level Items:

1.   Which of the following trends are expected to continue into the next decade?

   a.   More children are being born than are dying of malnutrition.
   b.   Global warming is decreasing.
   c.   The ozone layer is growing thicker.
   d.   The number of forests is increasing.

2.   Which of the following is(are) example(s) of the working poor in the U.S.?

   a.   the unskilled
   b.   displaced farm families
   c.   former blue collar workers forced out of their trades
   d.   a and b
   e.   b and c

3.   The term *sustainable* means the ability to continue indefinitely.

   a.   true
   b.   false

4.   Which of the following is a condition of uncertain or limited access to food of sufficient quality or quantity to sustain a healthy and active life?

   a.   hunger
   b.   food poverty
   c.   food shortage
   d.   food insecurity

5.   Which of the following is the leading cause of blindness in the world's young children?

   a.   vitamin A deficiency
   b.   iodine deficiency
   c.   vitamin C deficiency
   d.   protein deficiency

6. If you are a meat eater, which of the following recommendations would you follow in order to be environmentally conscious?

   a. Select canned corn beef.
   b. Choose chicken from local farms.
   c. Select small fish.
   d. a and b
   e. b and c

7. Which of the following fish species is endangered by overfishing?

   a. snapper
   b. flounder
   c. salmon
   d. shark

8. Which of these types of bags is the best choice from an environmental perspective?

   a. paper
   b. plastic
   c. cloth
   d. a and b
   e. b and c

9. Locally grown foods require less transportation, packaging, and refrigeration.

   a. true
   b. false

10. Which of the following packaging techniques for processed foods is the most environmentally advantageous?

    a. meat wrapped in butcher paper
    b. eggs in foam cartons
    c. juices in small individual cartons
    d. grain products in separate little packages

11. Which of the following cooking aids is the best choice from an environmental perspective?

    a. paper towels
    b. plastic wrap
    c. dishcloths
    d. aluminum foil

12. Refrigerators are one of the kitchen appliances that use the most energy in most people's homes.

    a. true
    b. false

13. In the United States, hunger is the result of:

    a. food insecurity.
    b. food shortage.
    c. famine.
    d. food poverty.

14. In the United States, one of every _____ Americans receives food assistance of some kind.

    a. 2
    b. 4
    c. 6
    d. 10

15. Hunger is not always easy to recognize.

    a. true
    b. false

16. To help solve the world's environmental, poverty, and hunger problems, the poor nations should:

    a. make contraceptive technology more widely available.
    b. educate their citizens.
    c. stem their wasteful uses of resources and energy.
    d. a and b
    e. b and c

17. Example(s) of the chronic poor include(s):

    a. migrant workers.
    b. the homeless.
    c. displaced farm families.
    d. a and b
    e. b and c

18. Recipients can use food stamps to purchase:

    a. alcohol.
    b. cleaning items.
    c. seeds.
    d. tobacco.

19. Food recovery from private industry has become a national priority.

    a. true
    b. false

20. Which of the following is a prerequisite to curbing population growth?

    a. relieving poverty
    b. inventing effective birth control measures
    c. alleviating hunger
    d. a and b
    e. a and c

21. Most children who die of malnutrition do so because:

    a. they starve to death from lack of adequate food intake.
    b. their mothers keep the food for themselves and their father.
    c. they lack adequate high-quality protein from animal sources.
    d. their health has been compromised by dehydration from infections that cause diarrhea.

22. The numbers of people affected by famine are relatively small compared with those suffering from less severe but chronic hunger.

    a. true
    b. false

23. The nutrients most likely to be lacking in malnourished individuals include:

    a. vitamin A, vitamin C, and iron.
    b. iodine, iron, and vitamin A.
    c. zinc, protein, and vitamin C.
    d. iodine, zinc, and vitamin D.

24. Population stabilization is one of the most pressing needs of our time because it appears to be the only way to enable the world's food output to keep up with demands.

    a. true
    b. false

25. Consumption of these fish helps to protect over-fished species:

    a. Alaskan halibut.
    b. swordfish.
    c. white sea bass.
    d. a and b
    e. a and c

Application Level Items:

26. Which of the following would you do to be an environmentally conscious food shopper?

    a. Buy bulk items with minimal packaging.
    b. Choose beef over fish.
    c. Eat canned chili.
    d. Choose fish which has been flown in from far away.

27. By purchasing fresh foods grown locally you are:

    a. selecting nutritionally superior foods.
    b. saving money.
    c. helping the local economy.
    d. minimizing negative impacts on the environment.

28. You want to cook a whole roast for dinner. Which of the following would be the safest and best way to save energy?

    a. Use the microwave.
    b. Use the oven.
    c. Use the outdoor gas grill.
    d. Use the pressure cooker.

29. To solve the world's environmental, poverty, and hunger problems, rich nations must:

    a. gain control of their population growth.
    b. stem their wasteful and polluting uses of resources and energy.
    c. reverse the destruction of their environmental resources.
    d. educate their citizens.

30. Which of the following steps should be taken to relieve poverty and environmental degradation worldwide?

    a. Ease the debt burden that many poor nations face.
    b. Adopt sustainable development practices that slow destruction of waterways, forests, and soil.
    c. Emphasize technology-intensive methods of harvesting the resources of developing countries.
    d. a and b
    e. b and c

Answers

Summing Up

(1) 13; (2) food insecurity; (3) food poverty; (4) working; (5) environmental; (6) soil erosion; (7) population; (8) poverty; (9) economic; (10) education; (11) hunger; (12) contraceptive; (13) educate; (14) sustainable; (15) energy; (16) economies; (17) sustainable; (18) environmental; (19) environment; (20) shopping; (21) food chain; (22) packaging; (23) fast; (24) appliances; (25) Energy Star

Chapter Glossary

Matching Exercise: (1) h; (2) c; (3) j; (4) f; (5) k; (6) a; (7) g; (8) b; (9) i; (10) e; (11) l; (12) d

Exercises

Chapter Study Questions:

1. The term *sustainable* refers to the use of resources at such a rate that the earth can keep on replacing them. An example would be cutting trees no faster than new ones grow.

2. Because growing animals for their meat and dairy products by feeding them grain uses more land and other resources than growing grain and other crops for direct use by people. In addition, animals use more water, add more pollution to waterways, and, in general, cause destruction of more native vegetation and wildlife than equivalent amounts of plants.

3. Fast cooking methods use less energy than conventional stovetop or oven cooking methods. Examples include stir-frying, pressure cooking, and microwaving.

4. Poor people often destroy the resources they need for survival, such as soil and forests. When they cut the trees for firewood or timber to sell, they lose the soil to erosion and, without these resources, they become poorer still. Thus, poverty causes environmental ruin and hunger grows from environmental ruin.

5. Poverty and hunger are correlated with lack of education which includes lack of knowledge on how to control family size. Also, families in poverty depend on children to farm the land, haul water, and care for adults in their old age. Many young children living in poverty also die from disease or other causes, in which case parents choose to have many children as a form of "insurance" that some will survive to adulthood.

6. Because in many nations women are the sole breadwinners in their household and because many women are also engaged in farming. Also, even starving women give what food is available to their children. Providing support to women also supports activities that improve the conditions in the community.

Short Answer Questions:

1. (a) hunger, poverty, and population growth; (b) loss of food-producing land; (c) accelerating fossil fuel use; (d) increasing air pollution; (e) atmosphere and climate changes, droughts, and floods; (f) ozone loss from the outer atmosphere; (g) water shortages; (h) deforestation and desertification; (i) ocean pollution; (j) extinctions of species

2. food poverty

3. (a) drought; (b) flood; (c) pests

4. (a) stir-frying; (b) pressure cooking; (c) microwaving

5. (a) keep it set at 37-40° F; (b) clean the coils and the insulating gaskets around the door regularly; (c) keep it in good repair

6. (a) make contraceptive technology and information more widely available; (b) educate their citizens; (c) assist the poor; (d) adopt sustainable development practices that slow and reverse the destruction of their forests, waterways, and soil

7. (a) health care; (b) family planning; (c) education

8. (a) tobacco; (b) cleaning items; (c) alcohol; (d) nonfood items

9. stem their wasteful and polluting uses of resources and energy

10. (a) migrant workers; (b) the unskilled and unemployed; (c) the homeless; (d) some elderly

Problem-Solving:

 X car pool to the grocery store
 __ eat high on the food chain
 X buy foods which are locally grown
 __ buy canned beef products
 __ select fish such as tuna, swordfish, and shark
 X buy items in bulk
 __ use plastic bags instead of paper ones
 X use the microwave for cooking

Controversy Questions:

1. Agriculture that is practiced on a massive scale by large corporations owing vast acreages and employing intensive technologies, fuel and chemical inputs.

2. By causing ozone depletion, water pollution, ocean pollution, and other ills and by making global warming likely.

3. Large masses of animal waste are produced and they leach into local soils and water supplies, polluting them.

4. Farmers using this system employ many techniques, such as crop rotation and natural predators, to control pests rather than relying on heavy use of pesticides alone.

5. By shrinking the acreage needed for crops, reducing soil losses, minimizing use of chemical insecticides, and bettering crop protection.

Sample Test Items

1. a (p. 561)	11. c (p. 574)	21. d (p. 567)
2. e (p. 563)	12. a (p. 575)	22. a (p. 567)
3. a (p. 572)	13. d (p. 563)	23. b (p. 567)
4. d (p. 561)	14. c (p. 564)	24. a (p. 570)
5. a (p. 563)	15. a (p. 563)	25. e (p. 569)
6. e (p. 574)	16. d (p. 572)	26. a (p. 574)
7. d (p. 569)	17. d (p. 563)	27. d (p. 574)
8. c (p. 574)	18. c (p. 564)	28. d (p. 574)
9. a (p. 574)	19. a (p. 564-565)	29. b (p. 572)
10. a (p. 574)	20. e (p. 570-571)	30. b (p. 572)